Encyclopaedia of
SOLID STATE PHYSICS

Encyclopaedia of SOLID STATE PHYSICS

Dr. Dinesh Bhaskar

ANMOL PUBLICATIONS PVT. LTD.
NEW DELHI-110 002 (INDIA)

ANMOL PUBLICATIONS PVT. LTD.

Regd. Office: 4360/4, Ansari Road, Daryaganj,
New Delhi-110002 (India)
Tel.: 23278000, 23261597, 23286875, 23255577
Fax: 91-11-23280289
Email: anmolpub@gmail.com
Visit us at: www.anmolpublications.com

Branch Office: No. 1015, Ist Main Road, BSK IIIrd Stage
IIIrd Phase, IIIrd Block, Bangalore-560 085 (India)
Tel.: 080-41723429 • Fax: 080-26723604
Email: anmolpublicationsbangalore@gmail.com

Encyclopaedia of Solid State Physics

First Edition, **2011**

ISBN 978-81-261-4233-0

PRINTED IN INDIA

Printed at Mehra Offset Press, Delhi.

Contents

Contents

Preface

Solid-state physics is the largest branch of condensed matter physics. It is the study of rigid matter, or solids, through methods such as quantum mechanics, crystall-ography, electromagnetism and metallurgy. Solid-state physics considers how the large-scale properties of solid materials result from their atomic-scale properties. Solid-state physics thus forms the theoretical basis of materials science, as well as having direct applications, for example in the technology of transistors and semiconductors.

Solid materials are formed from densely-packed atoms, with intense interaction forces between them. These interactions are responsible for the mechanical (e.g. hardness and elasticity), thermal, electrical, magnetic and optical properties of solids. Depending on the material involved and the conditions in which it was formed, the atoms may be arranged in a regular, geometric pattern (crystalline solids, which include metals and ordinary water ice) or irregularly (an amorphous solid such as common window glass).

The bulk of solid-state physics theory and research is focused on crystals, largely because the periodicity of atoms in a crystal its defining characteristic facilitates mathematical modeling, and also because crystalline materials often have electrical, magnetic, optical, or mechanical properties that can be exploited for engineering purposes. Many properties of materials are affected by their crystal structure. This structure can be investigated using a range of crystallographic techniques, including X-ray crystallography, neutron diffraction and electron diffraction.

Author

Preface

Solid state physics is the largest branch of condensed matter [illegible]

Chapter 1

Physics of Solids

We are all familiar with the matter that makes up the world and ourselves, as well as the fact that it comes in three different forms: as solid, liquid, or gas. This chapter discusses the properties of solid matter, with a focus on the applications of those principles to the design of structures. Liquids and gases are discussed in the next chapter.

MATTER AND ATOMS

The idea that matter could be broken down into submicroscopic individual units, known as "atoms", goes back to antiquity, though the concept didn't start to acquire scientific momentum until 1808, when the English scientist John Dalton proposed it to account for the results of his studies of chemical reactions.

Today, it is known that there are about a hundred different types of atoms with different weights and other properties. These "elements" include hydrogen, the lightest and most common element; carbon, which makes up soot; silicon, one of the constituents of sand; metals like gold, iron, and lead; the gaseous elements oxygen and nitrogen; and many others.

Atoms could also be combined to form "molecules". For example, if the element oxygen is burned with the element hydrogen, each oxygen atom forms "chemical bonds" with two hydrogen atoms to form water molecules.

It was the distinct and predictable patterns of combinations of atoms into molecules that led Dalton to propose the atomic theory. Large numbers of atoms can combine into a single molecule, particularly in the case of the complex carbon-based "biomolecules" that make up living organisms, and the materials made from atoms and molecules can be arranged into large-scale systems, from stones to machines to flowers to human beings.

Detailed discussion of chemical reactions, in which atoms are converted into molecules or molecules are broken down into atoms, needs to be reserved for a document on chemistry. However, it is useful to comment here that such chemical reactions observe the fundamental law of energy conservation as obeyed in physics in general. Reactions may be "exothermic", releasing energy stored in chemical bonds, or "endothermic", absorbing energy stored into chemical bonds; but in any case the total energy before and after the reaction will remain the same. Only the forms are changed.

For example, water can be broken down into hydrogen and oxygen by running an electric current through it; this is an endothermic process. A spark will then cause the hydrogen and oxygen to explosively recombine; this is an exothermic process.

On a broader canvas, a tree grows by using sunlight and raw materials such as water and carbon dioxide to drive its complicated biological processes, which in sum amounts to an endothermic process. The wood can be burned to release its energy, which is an exothermic process.

Oddly, although atomic theory now seems to be an essential element of classical physics, it wasn't completely accepted by physicists until the early 1900s. Ironically, this was at the very time that the doctrines of classical physics would begin to be seriously challenged.

Atoms had been originally thought of as indivisible, but even before the final triumph of atomic theory physicists were beginning to realise that atoms themselves were not indivisible and had an internal structure. They gradually realized that an atom had a heavy central "nucleus" that had positive electric charge, surrounded by light orbiting "electrons" that had a negative electric charge. Electrons could be stripped from atoms to be sent through wires as electrical currents.

If an atom lost its normal complement of electrons, it became a positively charged "ion". Atoms could also acquire an excess of electrons, becoming negatively charged ions. Such charge imbalances were one mechanism that contributed to the bonding of atoms into molecules, though subtler processes were involved as well.

Physicists soon found that the interior of the atom was a complicated place where things happened that were impossible at a larger scale, and this simple model of a "classical atom" had to be revised. The revisions lead into the realm of modern physics and so are beyond the scope of this document. The classical atom still provides a useful basis for the discussion of the fundamental properties of matter.

GROSS PROPERTIES OF SOLIDS

Solid matter covers a wide range of forms, everything from a compacted block of trash containing every form of household rubbish to a glittering diamond, consisting of carbon atoms arranged in an orderly, uniform "crystalline" arrangement.

At a high level of abstraction, chunks of matter have certain properties that can be characterized and measured, for example colour; density, or ratio of mass to volume; and various measures of strength, or the ability of that chunk of matter to retain its structure or "integrity" when subjected to various forces. Obviously these things are much easier to characterize for, say, a diamond than for a block of compacted trash.Solid matter also has certain properties of "scale" that are independent of what the matter is made up of. For example, let's take a set of cubes of solid matter without regard to what the matter is. Suppose one cube is a centimeter on a side. Now let's suppose Dexter stacks up the cubes to make a bigger cube that is two centimeters on a side.

This doubles the linear dimensions of the cube. The surface area of the cube has expanded by a factor of four, while the volume and mass of the cube has expanded by a factor of eight. In more formal terms, doubling the linear dimensions of an object squares its surface area and cubes its volume and mass. This means that as an object grows bigger, its volume and mass increase much more rapidly than its surface area.

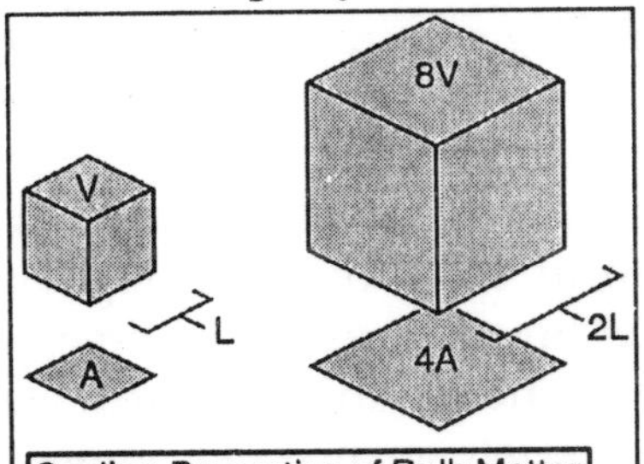

Scaling Properties of Bulk Matter

1950s horror movies such as Them envisioned monster ants, but in fact such scaling effects prevent insects from becoming very large. For one thing, they don't have lungs, and have to acquire oxygen through pores in their chitinous skins.

Doubling their size would increase their volume and mass twice as fast as it would increase their surface area, halving their ability to obtain oxygen for that mass, and at a certain size a giant insect would simply suffocate.A more important scaling issue that rules out giant insects is the issue of "compressive strength", or the ability of a structural support to bear weight placed on top of it.

The compressive strength of a structural support, such as a column that holds up a building, is proportional to the cross-sectional area of a horizontal slice through the column.A cross section is just one face of the surface of an object, and so like surface area increases with the square of a linear increase in size. However, the mass of the building increases with the cube of a linear increase in size.

Double the linear dimensions of a building and the ability of a column to support the building's weight is increased by four, while the mass and the load on the column is increased by eight. The column now has to proportionately bear twice as much load as it did at the smaller scale. This means that more, or disproportionately large, columns must be used to support a larger building.

Insects generally have spindly bodies and legs, while an elephant has great stumpy massive legs. If an insect was scaled up to large size, it would simply collapse of its own weight. However, horror movie fans can still take comfort that deadly venomous or parasitic insects are not ruled out by the laws of physics.

STRENGTH OF MATERIALS

These considerations of scale lead back to the notion of strength of materials, which in turn leads to consideration of how materials are used for building structures.

Although builders always had useful concepts of what could or could not be done when building structures, the first person to try to perform scientific experiments to determine the strength of materials was the Italian polymath Leonardo da Vinci.

The first formal text on the subject of materials was actually published late in life by Galileo in 1638, titled dialogues concerning two new sciences. He was under house arrest at the time, having antagonized the Vatican by promoting the notion that the Earth went around the Sun and not the reverse, and materials seemed like a noncontroversial subject not likely to get him into further trouble.

Galileo's dialogues was not particularly correct in all details the basic laws of physics needed hadn't been invented just yet but the book did get people thinking. Isaac Newton had little interest in the subject of materials, but his rival Robert Hooke was fascinated by the subject, examining the ways in which materials responded to forces, in particular coming up with Hooke's law. This was mentioned in a previous chapter as:

$$\text{Force} = \text{stiffness} \times \text{compression}$$

It was stated then as applying to a spring, but it also applies to structural element made of hard metals, stone, and glass. It does not in general apply to flexible materials like rubber or tendon, which stretch easily up to a limit but then yield diminishing returns. Hooke's work was the first to at least imply the notions of "stress" and "strain".

Stress is a force applied to a structural element. Such forces include:

- Compression forces, which squeeze the element in on itself.

- Tension forces, which try to pull the element apart.
- Shear forces, like we exert on a piece of cardboard to rip it in half.
- Torsion or twisting forces.
- Bending forces.

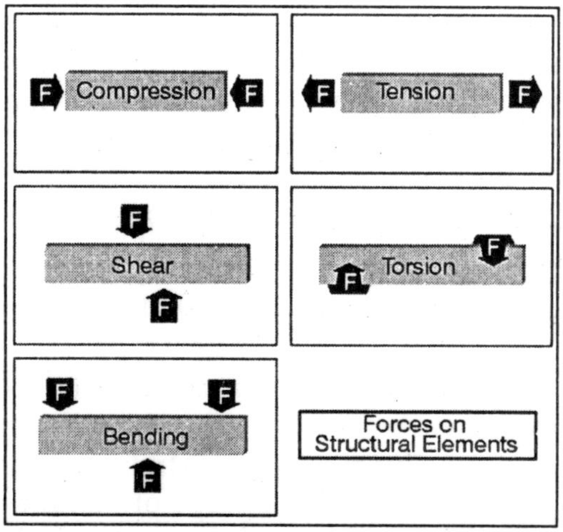

Stress is measured in terms of force per unit of area, or what is simply known as "pressure". The formal metric unit of pressure is the "pascal (pa)", which is a newton per square meter, named after the great French mathematician and physicist Blaise Pascal.

Since a pascal is a somewhat small unit, it is often expressed in terms of "kilopascals (kpa)" thousands of pascals or "megapascals (Mpa)" megapascals. In compression and tension forces, the force is applied directly through a cross-section of a structural element, and so the stress is measured in terms of Mpa.

In shear force, the force is applied along a plane through a structural element, and so stress is measured in terms of Mpa over the area of the interface between the two sections of the structural element. Torsion is similar to shear but involves a torque over the area of the interface.

Bending is complicated, since it involves a shear force down through the structural element, a tension force along its bottom, and a compression force along its top. In practice, structural elements may be subjected to a number of these stresses at the same time.

A structural element under stress will exhibit a "strain", that is, a lengthening, given in a simple proportion. Up to a certain limit, once the stress is removed, the structural member will return to its original length. This is called "elastic" behaviour".

It was certainly implied by Hooke, but the idea was developed further by Thomas Young, who published a paper in 1807 in which he defined the concept of "elastic modulus" or "Young's modulus". In modern terms, this is the ratio of stress of a particular material to its strain:

elastic modulus = stress/strain

It is somewhat more intuitive to express this as:

strain = stress/elastic modulus

Since strain is a simple proportion, the units of elastic modulus are still Mpa. For materials that stretch easily. the elastic modulus has a low value; for example, rubber has an elastic modulus of 7. The stiffer the material, the higher the elastic modulus:

Wood (spruce)	11,000 Mpa
Ordinary glass	70,000 Mpa
Steel	201,000 Mpa
Diamond	1,200,000 Mpa

Given these definitions of stress and strain, then Hooke's law can be rephrased as follows:

force = stress × strain

Hooke's law also leads to the concept of "strain energy", or the potential energy contained in a strained structural element, which is similar to the eq4uation for energy in a stretched spring:

strain energy = volume ×(1/2) × stress × strain^2

Once the stress on a structural member exceeds a certain limit, the member will be permanently strained out of shape, and will break if the stress is increased. Not all materials respond to forces in such a nice graded way. They may not deform at all up to the level of force where they snap. Such materials are said to be "brittle". Glass is a good example, since any practical experience shows how difficult it is to bend a glass rod without breaking it. Materials are said to have a "breaking strength" or "tensile strength", rated in the tension stress that causes them to snap. The concept goes back at least to Galileo, who performed tests in breaking strengths, though he didn't have the theoretical tools needed to describe it well.

Typical tensile strengths include:

High strength steel Mpa	1,500
Ordinary commercial steel Mpa	400
Cast iron	~200 Mpa
Aluminum alloys	~300 Mpa
Hemp rope	80 Mpa
Bone	40 Mpa
Nylon fibre	1,000 Mpa
Kevlar fibre	2,700 Mpa

Not all materials respond to different forces in the same way. Although metals are strong in both compression and tension, stone is very strong in compression but is weak in tension. Wood can be split easily by shear forces applied along the line of its grain, but will resist larger shear forces applied at a right angle to its grain. Forces acting consistently on a structural element over a long period of time may cause a slow deformation, or "creep", that can lead to failure even if the elastic limit is never exceeded. Similarly, a cycle of changing stresses repeated over and over may cause "fatigue" that will eventually lead to the failure of a structural element.

Although Hooke got the ball rolling, the notions of stress and strain were not articulated in their modern form until 1822, in a document published by a by a French mathematician named Augustin Cauchy. It would take time for Cauchy's work to sink in, since he wrote in a theoretical style that was more or less foreign to engineers, but now his concepts are regarded as one of the major foundations of structural design.

The underlying idea of all structural design is balance of stresses. If the stresses imposed on a structural element exceed the strength of the element, it will collapse. The stresses are set up by what engineers refer to as "loads". Some of the loads on a structure are obvious: the "dead load" set up by the weight of the structure, and the "live load" set up by the people, furnishings, equipment, and materials contained in the structure. These are known as "static loads" because they don't change, or don't change much, over time.

Others loads are not quite so obvious. There are the "hidden" loads, with stresses set up by the expansion or contraction of structural elements due to changes in temperature, or by unequal settling of the building's foundations. Structures may also be subjected to rapidly changing "dynamic" loads, such as those caused by winds and earthquakes.

Dynamic loads can be particularly treacherous since they are somewhat unpredictable. In one famous example, the designer of the Citicorp skyscraper in New York City received a letter from a student who was curious about the building's stability in winds. The architect replied that the building's design had been analyzed to ensure that it would stand up to storm winds and that it was safe.

However, after sending his reply, the architect realized that the analysis had been performed with the assumption that the winds would be perpendicular to the faces of the building. He re-did his analysis to see what would happen if the winds were at an angle, and to his shock found out that a severe windstorm would generate torsion forces that would bring the skyscraper down in the middle of Manhattan. The Citicorp building's steel framework was hastily reinforced.

DESIGN OF STRUCTURES

Structures can be divided into two classes: compressive structures, in which the structural elements are stacked on top of each other; and tensile or "suspension" structures, in which the structural elements are suspended from supports. One of the simplest examples of a compression structure is a brick wall, with the bricks piled on top of each other. One of the simplest examples of a tensile structure is a circus tent, with the canvas "big top" held up by ropes hung from a central pole. Of course, in practice most structures feature both compressive and tensile elements.

The first really big structures were compression structures, since it was difficult to build durable large structures of wood, and stone doesn't lend itself to tensile structures. The major problem with stone compressive structures is that stone is very heavy, or more specifically has a low "strength to weight" ratio.This means that the weight of such a structure grows very quickly as it the structure grows bigger, and walls have to be made thicker and thicker at the base so the structure doesn't collapse. This approach ultimately leads to the simplest of all large stone structures, the pyramid, built not only in Egypt but in China and in the Pre-Columbian New World as well.

Pyramids are an impressive but purely brute-force design. Builders were also learning more sophisticated tricks working with wood, the other classic building material. Wood was used to construct ordinary buildings, such as homes, and structures such as bridges.

The earliest wooden bridges no doubt consisted just of two beams set up across a stream and covered with planks. The major problem with this is that such a bridge is limited in length. It is supported at the ends but not in the middle, and as it gets longer so does the "lever arm" of the half-spans of the bridge, meaning that a person stepping on the middle of the bridge places a proportionately greater stress on it. If the bridge was over the top of a shallow stream, this problem could be addressed by driving pilings in the middle of the stream and fastening the middle of the span to them, but if the stream wasn't shallow or if the bridge was over the top of deep canyon, such a fix wasn't possible.

There were two possible solutions. The first was the rope "suspension bridge", a tension-based design. Wooden or stone towers could be constructed on both sides of the canyon, with heavy ropes slung across their tops and anchored to the ground. A wooden walkway could then be suspended from the ropes. This type of bridge has been around for millennia, and is thoroughly familiar from B-fiction adventure movies since they are excellent sites for dramatic action scenes. They always seem to be in a frighteningly bad state of repair, and the waters beneath are often full of crocodiles, piranhas, or other predatory creatures.

For present purposes, a second option is more interesting. Suppose Dexter has a wooden bridge and wants to brace up the middle of the span from each end. He can do this by fitting a beam from the bank to the midspan at an angle.

This allows the forces directed downward at midspan to be shunted towards the banks. Since this sets up a sideways force on the two support beams that tends to splay them apart from the middle, it is also useful to link them together at the bottom with a third beam, forming a triangular support structure known as a "truss". The truss is a basic structural form, and is composed of various triangular structural subunits linked together.

The triangular shape is very useful as it will stand up to stresses imposed in different directions, and it is also employed in the "cross-bracing" of structures used to keep them from falling over in heavy winds and the like. The truss is of course in as wide use as it ever was; one of the tidiest examples is the metal high-voltage transmission line tower, which is a lightweight structure made up of little triangular truss elements.

The problem with wooden bridges is that they didn't tend to last very long. Stone is far more durable, but it does not lend itself to construction of trusses. However, the same underlying principle of diverting stresses to the sides could be used on stone bridges as well, by building them with curved structures called "arches".

The arch did imply the provision of supports to deal with the sideways loads it set up. This was generally simple with bridges, as they might be built between stony riverbanks or, if such weren't available, heavy footings could be constructed on each bank.

This could be a tricky matter for other structures. The Gothic cathedrals of Europe, with their high arched or "vaulted" ceilings, were built with "flying buttresses", structures attached to the outside of the building to provide the necessary lateral support. The "dome" was an extension of the arch, rotated around in a full circle.

Like the arch, the dome required lateral support to handle the horizontal stresses. Islamic architecture, which demonstrates a certain fondness for domes, often involves half-domes or "semi-domes" ringing the main dome to distribute sideways forces down to the heavy walls of the buildings.

Although complete spheres are not ordinarily used in buildings, they are an extension of the notion of an arch or dome, in which compression is distributed around the sphere. An egg can be surprisingly hard to crack if subjected to balanced compression forces try carefully squeezing a hard-boiled egg and spheres are often used in the crew-carrying elements of deep-ocean submersibles, since they can withstand extreme pressures.

Most basic elements of classical building design, like the arch and dome, were in place by Roman times. Major advances in structural design had to wait until the development of large-scale metalworking in the 19th century. Iron and then steel structural elements, with their strength in both compression and tension and their high strength-to-weight ratio, made possible great steps forward in building design.

Powr Tower (GVG/PD)

Classical construction forms, such as trusses and arches, could be built with such metal beams. The beams were generally in the form of "I-beams", which consist of a length of metal with a long flat horizontal plate on the bottom, a similar long flat horizontal plate on top, and a vertical plate connecting the two horizontal plates. A beam is subjected to compression on top and tension on the bottom, and the I-beam is designed to provide the maximum cross-sectional area on bottom and top to resist these stresses. Such techniques could be used to build new sorts of structures, such as the modern skyscraper. The poor strength-to-weight ratio of stone made very tall buildings impossible, but metal structural elements allowed buildings to be made much taller. A skyscraper consists of a "cage" of steel girders that supports the floors and the walls.

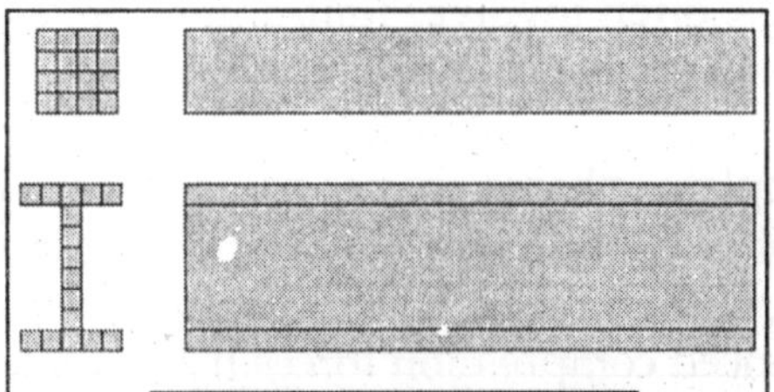

SIMPLE BEAM VERSUS I-BEAM

The spectacular explosive demolition of such structures, which usually makes local or even national TV news, is based on the sequential detonation of charges that cut critical structural supports of the cage in a particular order, causing the building to cave in on itself in a heap if the event has been properly planned.

Metal building elements also allowed the development of large truss bridges, with a latticework of metal beams to allow distribution of stresses. It also led to the development of modern suspension bridges in which the roadbed is hung from huge cables slung over high towers and anchored to heavy concrete "anchors" at each end and "cable-stay" bridges in which the roadbed is held up by cables connected directly to the towers.

Concrete was around in Roman times, though the recipe was lost in the Middle Ages and not rediscovered until the 19th century. Modern building techniques make it far more useful than it was to the Romans. Using steel reinforcing rods, it is commonly used for foundations, walls, and other structural elements under compressive loads.

However, "prestressed" concrete elements can also be used to handle tension loads that would cause ordinary reinforced concrete to crack. A prestressed concrete structural element is cast in a frame that supports steel cables under tension and laced through the concrete. When the concrete hardens or "cures", the frame is removed, and the tension of the cables within the concrete matrix prevents it from cracking under tension loads.

There is a balance in the design of structures in the number of structural elements required. Reducing the number of elements reduces complexity, generally reducing the labour of assembly and cost. However, reducing the number of elements also usually means larger elements, which may be difficult to handle, and also more prone to damage.

Suppose a dome is made up of curved panels; the fewer the number of panels, the larger the panels, the greater the load each panel has to carry, making each panel more prone to cracking and also making replacement of the panels more troublesome.

Ofcourse, if there are a very large number of small panels, installing them will be laborious, and trying to make sure the dome doesn't leak in the rain through the seams between the panels may be difficult. There are cases where the benefits of "subdivision" are obvious. The cables used to hold up a suspension bridge are each made of large numbers of individual strands; the breakage of a few strands has little effect on the strength of the cable as a whole. It is also possible to construct structural elements as "laminates", using layers of materials that are bonded together with adhesives.

Plywood is a common example of a laminate, and panel structures may be made as "sandwiches", with a lightweight core and stiff surface paneling, providing an improved strength to weight ratio. Simple corrugated cardboard is a common example of such a sandwich material; much the same approach can be used with metal panels as well, with a corrugated or honeycomb fill. However, laminates do present the danger of becoming separated or "delaminated" under stress.

Similar principles applies to the types of "composite" materials that consist of threads of one material embedded in a matrix of another material. Examples include glass or graphite fibers embedded in plastic, or "whiskers" of silicon carbide embedded in metal.

Any failure of a few threads does not do much to undermine the integrity of the structural element made of the composite, and the threads prevent the propagation of cracks. Research is now being conducted on also incorporating sensors in composites in the form of threads to give indication of when a structural element is about to fail, but such "smart" materials are not in common use yet.

Chapter 2

Physics of Liquids and Gases

This chapter follows up the discussion of the physics of solids by providing a survey of the physics of liquids and gases.

BUOYANCY

Liquids have some properties similar to those of solids, such as colour and density, but of course they differ in their ability to flow and to take the shape of any container they are poured into. They possess "viscosity", which is a measure of their ability to flow. A "thick" liquid that is hard to shove through a pipe has a high viscosity, while a "runny" liquid that flows quickly through a pipe has a low viscosity.

It is obvious that it is much easier to get vinegar out of a squeeze bottle than honey. Viscosity is related to the concept of "fluid friction": concepts of friction and drag apply to liquids, and a liquid with a high viscosity generates a high level of fluid friction.

Another obvious characteristic of liquids is that a body of liquid will seek a common level. A pond undisturbed by wind may be "smooth as glass", as the saying goes, and it will be perfectly level, at least if it's small relative to the curvature of the Earth. It has to be noted here that in more general terms, the angle of the surface of a liquid will be at a right angle to the direction of the force applied to a liquid, so water in a spinning bucket will tend to "climb up the walls" of the bucket. An undisturbed pond will be flat because it is only subject to the pull of gravity, which is straight down.

If pipes of various diameters are stuck through the surface of this flat pond, it only takes a moment's thought to realise that the water level is the same no matter the diameter of a pipe.

Yet another obvious property of liquids is that not only can they flow through pipes, objects can flow through or over them. This leads to the concept of "buoyancy", the tendency of objects to float or sink in a liquid.

This is unsurprisingly due to the density of the object relative to the liquid. A block of wood is less dense than water and floats on water, while a rock is more dense than water and sinks.

The sunken rock will displace a volume of water equal to the volume of the rock. The floating block of wood has a slightly subtler behaviour. It will sink into the water under the influence of gravity until it displaces a volume of water equal to its own weight. At that point, it will reach an equilibrium of forces and sink no further.

This is known as "Archimedes' Principle", since it was devised by the Hellenic scientist Archimedes . The forces supporting this floating block of wood will focus upward and inward on it, summing to a point known as the "centre of buoyancy". The centre of buoyancy is always above the centre of mass of a floating object; the farther the distance between the centre of buoyancy and the centre of mass, the more stable the object that is, the less its tendency to capsize.

According to tradition, Archimedes discovered this fact while getting into a bathtub, and used it to determine for the authorities if a crown being sold by a metalsmith was actually made of gold or a debased metal. Archimedes weighed the crown, then determined its volume from the amount of water displaced, and from that determined its density, which could be compared with that of pure gold.

The story says the metalsmith was found guilty and executed, but the tale is probably a myth: it is a bit unlikely that Archimedes, using the instruments available to him, could have performed the required measurements with enough accuracy to have detected a fraud unless the debasement was so gross that it would have been obvious by simply inspecting the crown.

One simple example of the practical application of Archimedes' principle is the "Plimsoll line", which is a marking painted on the hull of a vessel in the form of a circle or diamond cut by a horizontal line. If the ship is sitting so low that the Plimsoll line is underwater, the ship is overloaded.

Another interesting example is provided by the question: if all the ice at the North Pole melted, how much would the sea level rise? From a simple consideration of buoyancy, the answer is: not at all. The polar ice displaces a mass of water equivalent to itself, and so when it melts, there is no overall change in the water level.

Of course, the ice melting off of Greenland and Antarctica would raise the sea level, and there's also the consideration of the extremely unpredictable ways that an ice-free Arctic Sea might affect global climate.

Incidentally, if a more dense liquid is poured into a less dense liquid, as long as the two liquids don't have a tendency to mix, the two will separate, with the less dense liquid on top and the more dense liquid on bottom. The classic example of this is oil and water.

Considerations of buoyancy govern the operation of a submarine. A submarine cruising on the surface of the ocean has an average density somewhat less than that of water. To submerge, it has to take in water in "ballast tanks", increasing its density to more than that of water and causing it to sink. To rise again to the surface, it has to purge the ballast tanks by blowing air into them, forcing the water out and lowering the submarine's density.

Once underwater, a submarine needs to maintain a density comparable to that of water so that it will not tend to either rise or sink. This issue is somewhat complicated by the fact that the submarine's hull is compressed as the vessel dives deeper, reducing the volume of the hull and so increasing the submarine's density.

In any case, once the submarine reaches the desired cruise depth, it can then maneuver freely by essentially "flying" underwater, using the "diving planes" or "foreplanes" that are mounted on nose or the "sail" (what's sometimes called the "conning tower") of the submarine, along with the "tailplanes", which are generally arranged in a vertical cross pattern.

However, there is an added complication for a submarine moving underwater. It is not firmly supported, and so if its centre of gravity shifts from the middle of the vessel, it will tend to tip down or up, making it difficult to control. This issue, known as "trim", was a particular problem for early submarines. It was solved by adding an additional set of "trim tanks", in addition to the ballast tanks.

There is a trim tank in the forward section of the submarine and a trim tank in the rear section of the submarine, with trim maintained by adjusting the relative amounts of water in these tanks.

A MODERN SUBMARINE

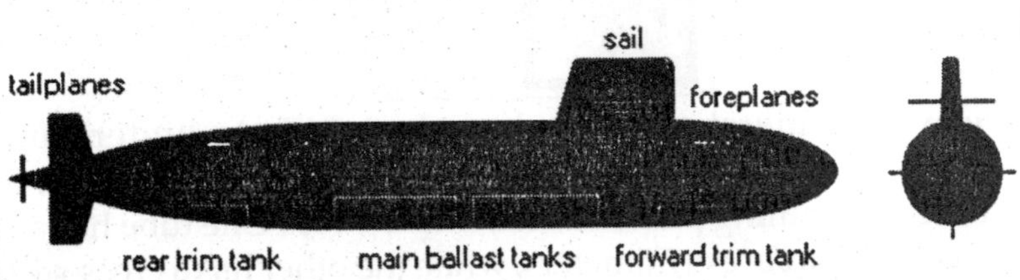

It takes a bit of time to pump water to shift trim, and so in World War II submarine warfare movies there is often a scene where the sub has to crash-dive, and all the crew that can be spared dash forward in a mad fire drill through the hatches to make the submarine nose-heavy.

LIQUIDS: FLUID PRESSURE/SURFACE TENSION

Although it is not intuitively obvious, the buoyancy of the submarine does not change in any significant way, no matter how deep it goes. Water is essentially incompressible and its density is effectively the same at any depth. Drop an iron block into the ocean and it will continue to sink at its terminal velocity a concept that applies in liquids just as it does in gases all the way to the bottom.

However, there is a limit to how deep a submarine can go, which is imposed by "water pressure", or the weight of the water above the submarine, which increases linearly with depth. Doubling the depth doubles the pressure.

There's nothing particularly tricky about this idea. Suppose Dexter has a set of identical concrete blocks, and proceed to stack them up one by one on top of each other.

Of course, the force per unit area on the bottom of the stack grows linearly with each block that is added. There is no difference if he stacks up containers of water instead of concrete blocks, or for that matter adds uniform increments of water to a single tall container.

What is not quite so intuitive is that the water pressure is the same in all directions. Suppose Dexter applies pressure to the top of a tank of water. Then, ignoring the increment of pressure added by the water in the tank as it gets deeper, the pressure of the water against the sides and bottom of the tank is exactly the same as the pressure on the water on top of the tank. This is known as "Pascal's principle", Pascal having published the concept in 1647.

One way to understand Pascal's principle is to understand the operation of a hydraulic jack. Suppose Dexter has a tank filled with water, with two tubes containing pistons fitted into the top. One tube has a cross-sectional area of 10 square centimeters, while the other has a cross-sectional area of 100 square centimeters.

If he exerts a downward force on the piston in the small tube, the transmission of water pressure on the piston in the big tube will result in an upward force ten times greater.

In other words, this arrangement gives Dexter a mechanical advantage of 10:1. As with levers and screw jacks and so on, of course he doesn't get something for nothing. A simple consideration of the relative volume of water moved from the small tube to the big tube in this example shows that the piston in the big tube only moves a tenth of the distance upward that Dexter shoves the piston in the small tube down.

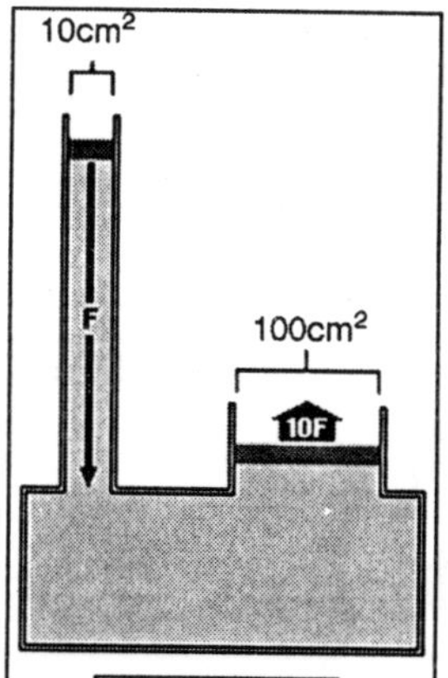

Hydraulic Jack

Another aspect of Pascal's principle is that it accounts for the buoyancy of a floating object, as discussed in the previous section. It is the force of pressure, exerted upward against the floating object, that prevents it from sinking further. Once the volume of water displaced equals the weight of the object, the force of water pressure will then support that object.

In the definition of Pascal's principle above, the assumptions were that the surface of the tank was under heavy pressure and the tank was not very deep, meaning the pressure was the almost the same at the top and bottom of the tank and the difference could be ignored.

For water of any real depth, the increment of pressure caused by the accumulated weight of water as the depth increases is of course not negligible. Water has a mass of a tonne (1,000 kilograms) per cubic meter, and so on Earth, where the gravitational acceleration is 9.81 meters per second, a cubic tank of water a meter on a side exerts a force of 1,000

9.81 = 9,810 newtons on the square-meter base, or 9.81 kilopascals. If Dexter keeps the sides of the base the same size, but doubles the height to two meters, the pressure grows linearly to 19.6 kilopascals. If he goes up to three meters, the pressure goes up to 29.4 kilopascals.

The water pressure on the sides of the tank increases linearly from zero at the top to the same as the bottom when it reaches the bottom.

However, let's suppose Dexter has a tank of water that's a meter high, a meter wide, and three meters long. This tank has a volume of three cubic meters. Is the bottom then under 29.4 kilopascals of pressure?

No, of course not. Pressure is force per unit area. He's increased the volume by a factor of three, but he's also increased the area of the base by a factor of three, and so the pressure remains the same as it was for the one-cubic-meter tank. The pressure is due to the weight of water pressing down only in the vertical direction. This means that it doesn't matter what size the base of the tank is. As long as the water in it is a meter deep, the pressure on the bottom in 9.8 kilopascals. This is more or less intuitive.

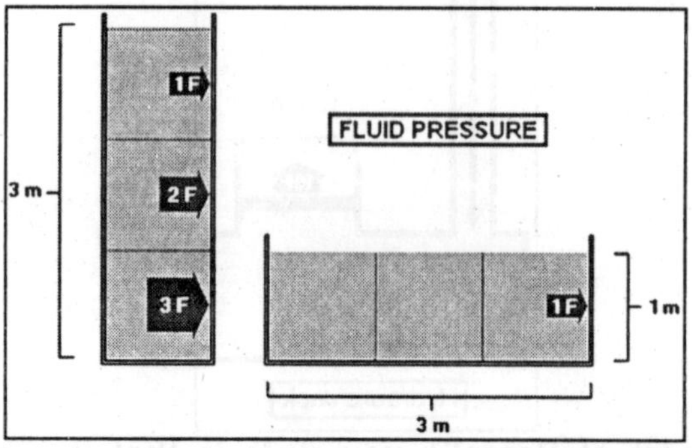

What is not so intuitive is that, as dictated by Pascal's principle, water pressure is exerted in *all* directions, and so a dam has to be just as strong to hold back a body of water that extends, say, only 10 meters behind the dam, as it must be to hold back a body of water of equal depth that extends 100 kilometers behind the dam. The depth is the only real determinant of fluid pressure.

There are two basic forms of dams: "gravity" dams, which simply use great mass to hold back the water, and "arch" dams, which are built in narrow rocky canyons in the form of a horizontal arch, which transfers the massive load of the water to the canyon walls. Some dams may use both mass and an arched configuration to ensure structural stability.

Obviously, the pressure of a liquid is equivalent to potential energy. If Dexter installs a valve at the bottom of a tank and opens it, then the water will pour out in a stream and will be able to do work, for example by turning a water-wheel.

This should not be surprising. A solid mass held at a height has potential energy due to its elevated position in a gravitational field, and the only difference between that and a liquid is that the liquid is in effect a continuous series of masses held at a continuous range of heights.

This makes calculation of the precise value of potential energy of the liquid in a tank a bit tricky. We won't do that here, but the important thing is to realise that water pressure is equivalent to potential energy, in the same way that the tension of a compressed spring is equivalent to potential energy.

Now let's suppose Dexter decides to tap this potential energy by opening the valve on the bottom of the tank. Water floods out in a stream. Since the water is now moving, it has kinetic energy. Of course, the higher the pressure at the valve, or equivalent the greater the height or "head" of water above the valve, the greater the kinetic energy of the flow and the greater its velocity. This is known as "Torricelli's principle", after the Italian physicist Evangelista Torricelli.

Since conservation of energy says that Dexter can't get something for nothing, this increase in kinetic energy must be obtained by a decrease in potential energy. As water pressure is equivalent to the potential energy of a liquid, this means that the pressure must decrease as well.

In other words, the faster a liquid moves, somewhat counterintuitively the more its pressure decreases. However, it's not so different in concept from Dexter firing a ball out of a spring-loaded toy gun: the spring tension is released and converted into the kinetic energy of the ball.

The same effect is observed when water or another liquid passes from a pipe with a wide cross-section into a pipe with a narrow cross section. For a given rate of liquid flow, measured in liters per second, the velocity of the flow, measured in meters per second, has to increase if the pipe diameter decreases to maintain the flow. In mathematical terms:

Liquid flow rate =
Cross section1 × velocity1 =
Cross section2 × velocity2

The increased velocity of the liquid flow in the narrow section of pipe will result in a decrease in liquid pressure. The following illustration demonstrates this principle, which is known as the "Bernoulli effect", after the Swiss physicist and mathematician Daniel Bernoulli.

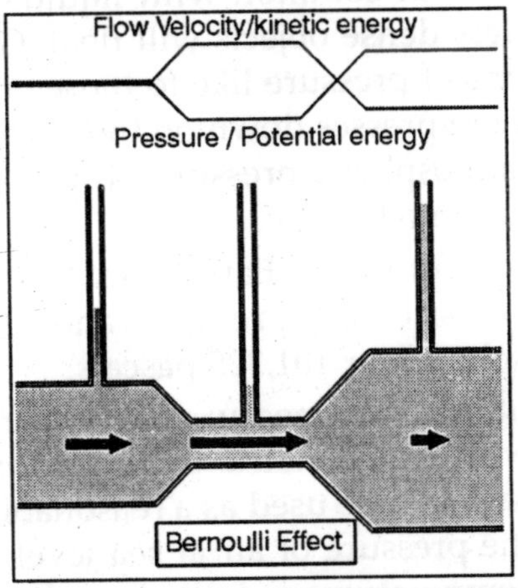

This illustration shows the pipe as horizontal. Of course, if it weren't, the liquid would flow "downhill" and pick up additional kinetic energy by the simple act of falling under gravity, complicating the analysis.

On a small scale, liquids can demonstrate behaviour that would seem bizarre on a larger scale, due to attractions between the molecules of the liquid.

For example, people can't walk on water, but insects like water-striders can. The intramolecular forces set up by water create "surface tension" that prevents a light object from simply falling right through the surface, even if it's denser than water.

Similarly, the adhesion of water to, say, glass surfaces will cause it to crawl up into glass tubes stuck into the surface of the water. This sort of "capillary action" is greater for tubes with very small diameters than it is for tubes with very large diameters.

Again, this is due to a scaling effect, since doubling the diameter of a tube only doubles its circumference and the area of its glass walls, while the cross-sectional area and the amount of water that is to be lifted increases by four times.

One interesting application of these phenomena is the "restartable" rocket engine. A rocket engine that has to be restarted in space has to the deal with the problem that in free-fall the propellants won't flow down out of the tank to the engine, since there is no "down".

There are various ways to deal with this problem; one ingenious scheme relevant to the current discussion is to fit the fuel tank with a set of internal grids of finer and finer spacing towards the bottom of the tank. The grids set up a "wicking" action that pulls the propellant to the bottom of the tank.

GASES

- Gases have much in common with liquids. They can flow, for example, and less dense objects will float. Gases also are subject to considerations of pressure like to those of a liquid; the weight of the atmosphere presses down on those who live at the bottom of it, and this atmospheric pressure drops off as we go to higher ground or fly up into the sky.

The atmospheric pressure of the Earth at sea level makes a convenient reference point for discussions of pressure, and one older measure of pressure is the "atmosphere", or 101,325 pascals.

This is a somewhat arbitrary measure of pressure, however, and not convenient to work with using metric units, but the value of 100,000 pascals, known as a "bar", can be used as a reasonable approximation, all the more so because the pressure of air at sea level actually varies with changes in humidity, temperature, and wind conditions.

A "barometer" is used to measure atmospheric pressure. A traditional scheme for building a barometer involves a U-shaped tube partially filled with mercury. One end of the tube is sealed and the other is open to the atmosphere. Mercury is squirted into the sealed end of the tube while it's upside down, and when the tube is turned over, the mercury falls down to the bend in the tube, leaving behind a vacuum in the sealed side.

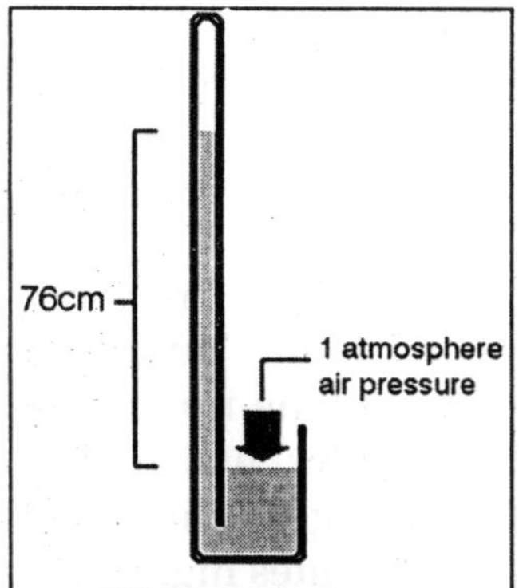

Mercury Barometer

Air pressure will push the mercury up into the vacuum, and the height of mercury on the sealed side will provide an indication of atmospheric pressure. The mercury will be 76 centimeters higher in the sealed side of the tube than the open side of the tube at a standard atmosphere of pressure. Of course, since pressure is force per unit area, it doesn't matter how wide the tube is, as long as its diameter is constant. This scheme was discovered by Torricelli in 1643, and a now-disused measure of pressure named the "torr", equivalent to the displacement of a millimeter of mercury under standard conditions, was defined in his honour. Naturally, a standard atmosphere of pressure is equivalent to 760 torr, or 76 centimeters of mercury.

A more modern approach is to use a metal bellows that has been evacuated. The bellows is connected to an indicator needle through some mechanical arrangement that amplifies any minor motion in the bellows, and as the bellows flexes under changes in pressure, the needle moves to give the pressure value on the indicator scale.This device is known as an "anaeroid barometer". If the indicator scale is calibrated to give altitude instead of air pressure, the device is called an "altimeter".

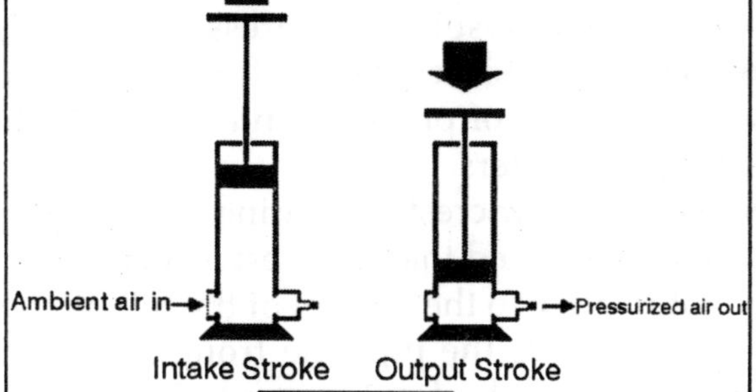

Bicycle Pump

Manipulating the pressure of air or other gases is the basis of "pneumatic" technologies. The most basic pneumatic device is the hand-driven air pump or "bicycle pump" as it's sometimes called, which involves a cylinder with a piston that is moved up and down by muscle power.

There's two valves in the cylinder, an "input" valve that opens inward under a pressure difference and an "output" valve that opens outward under a pressure difference. Pulling up the handle on the pump creates lower pressure inside the cylinder, which causes the input valve to pop open, with ambient air entering to eliminate the pressure difference. Pushing down on the handle creates higher pressure inside the cylinder, forcing the input valve closed and popping the output valve open to push air into the tire or whatever.

The notion of the mercury barometer leads to the concept of the "siphon", in which a liquid such as water can be driven from a reservoir to an outlet through a pipeline over a hill or other obstruction.

This vertical diversion is called a "siphon", and it works as long as two conditions are obeyed: the outlet has to be lower than the surface of the reservoir, or in other words water will still only flow downhill overall, and the siphon has to be filled with water before the scheme will work.

This looks like a problem in hydraulics, and it is, but the principle behind it is the same as that behind the mercury thermometer: the column of water flowing up the siphon is supported by the air pressure pressing down on the surface of the reservoir. When water flows out the outlet side of the siphon, it creates lower pressure, and so water on the inlet side of the siphon is driven upward.

That's why the siphon has to be filled before the siphon effect will work: if the outlet side of the siphon was open to the air, the water level in the input side of the siphon would be exactly the same as it was in the reservoir.

There is a limit to the height or "crest" of the siphon above the reservoir level, depending on the density of the liquid and the air pressure above the reservoir.

As mentioned, one atmosphere of pressure will lift mercury to a height of 76 centimeters, and since mercury is about 13.5 times more dense than water, one atmosphere of pressure on a reservoir should lift water about $13.5 \times 0.76 = 10.26$ meters.

This seems like a shallow crest for getting water over the top of a tall hill, but it must be remembered that the crest is relative to the water level in the reservoir, not relative to the bottom of the hill. Water reservoirs are often sited on high ground; the pipeline from the reservoir may drop down into a valley and then back up over a hill, but the crest of the hill will be about the same height or lower than the reservoir level.

Buoyancy issues are as relevant ın gases as they are in liquids. For example, if a balloon is filled with a light gas such as hydrogen or helium, it will have a lower density than the ambient atmosphere and tend to float up.

There is also the phenomenon of "lift" in heavier-than-air machines like aircraft and helicopters. There are misconceptions on how lift actually works, the biggest being that it is due to the Bernoulli effect, an error that actually pops up in elementary physics books.

As the story goes, the wings of an aircraft have a curved cross section, with a greater curvature on the top than on the bottom. Air flowing over the top of the wing has a longer path, and so its flow over the wing is more rapid. This means that the pressure on top of the wing is less than that on the bottom, and the effect is to provide a net upward force known as "lift" that keeps the machine flying.

The problem with this scenario become immediately obvious when the question is asked: "So how can aircraft fly upside down?" They don't normally, since it's uncomfortable for the pilot, but any good flight demonstration team like the US Air Force Thunderbirds flies their jet fighters upside-down as a regular practice and the machines aren't immediately shoved into the ground.

The fallacy with the Bernoulli effect argument is that it assumes the airflow over the top of the wing has to maintain mass flow with the airflow over the bottom of the wing when it isn't constrained to do so. The Bernoulli effect may have some effect on lift, but it's not the primary mechanism for the phenomenon.

The actual reason for lift is Newton's third law of motion. In flight, a wing is held at an "angle of attack (AOA)" upward of the direction of airflow, with the airflow striking the bottom surface and being deflected downward as "downwash".

The downward velocity component of the downwash provides a reaction to force the wing (and the aircraft connected to it) upward. The need to maintain AOA relative to the airflow does not mean the aircraft has to fly nose-up all the time; the wing may be fixed to the fuselage at a predetermined AOA relative to the aircraft's flight axis.

As far as this description goes, a simple flat wing would work as well as any other, so why is a wing typically built with a curved upper surface? The trick is that such a wing would produce a good deal of turbulence over the top of the wing, causing "buffeting" as well as some "upwash" of the airflow, which would tend to force the aircraft back down.

Due to some subtle features of fluid flow discussed in a bit more detail below the airflow tends to hug the curve of the upper surface, which means that it leaves the trailing edge of the wing in a downward direction, also contributing to lift.

The higher the AOA of the wing, incidentally, the greater the drag, and so there's a tradeoff between AOA and speed. It is not unusual for aircraft to have a landing gear configuration that leaves them nose-high, increasing the AOA until the aircraft gets off the ground and starts flying level. A high AOA is particularly important for carrier takeoffs and landings, since the aircraft has to get up or down in a short space.

A number of British carrier jets of the 1950s and 1960s were actually launched off carrier catapults strapped down so their nosewheel was in the air to ensure a high AOA. The US Navy Vought F-8 Crusader fighter had a wing that hinged upward for takeoffs and landings to permit a high effective AOA while giving the pilot a good view forward.

Such considerations of lift show that a conventional aircraft, even though it can in principle fly upside down, isn't as efficient in flight when doing so, having to maintain a nose-up attitude to ensure a usable AOA. Stunter aircraft may have no fixed AOA and wings that are curved on both sides.

An aircraft has "flight control surfaces" that modify the airflow to control the machine's direction. There are moving panels at the trailing edge of each wing known as "ailerons" that are pivoted in an alternating fashion on one side up, on the other down to allow the aircraft to "roll".

There are "elevators" at the trailing edge of the horizontal tailplane that are pivoted together to cause the aircraft to "pitch" up and down. Finally, there is a "rudder" at the trailing edge of the vertical tailplane that can be pivoted to cause the aircraft to "yaw" back and forth.

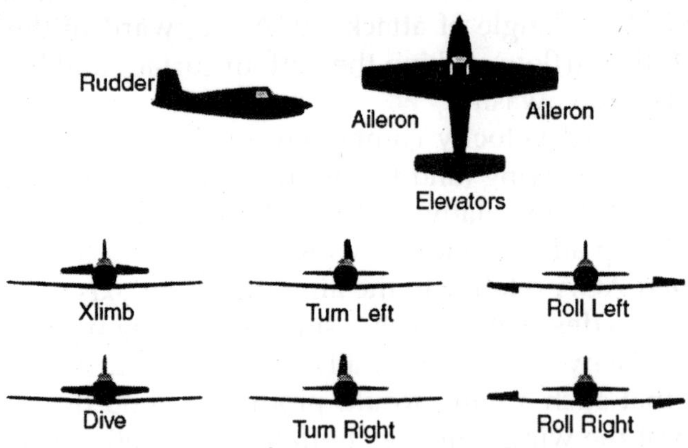

Aircraft Flight Control

These are the "basic" control surfaces, but there are many variations. For example, many tailless aircraft, usually delta-winged machines, have an "elevon" on each wing, with both moved up or down to provide elevator pitch control, or moved in opposite directions to provide roll control.

Tailless flying wing aircraft that don't have a tailfin, such as the B-2 Stealth bomber, have "rudderons" that are split top and bottom on the outer rear edge of the wing.

Both rudderons can be opened up top and bottom on one side to cause the aircraft to turn, or can be opened up top and bottom on both sides of the aircraft to act as airbrakes. An elevon is fitted on each half of the wing inboard of the rudderon to provide pitch and roll control.

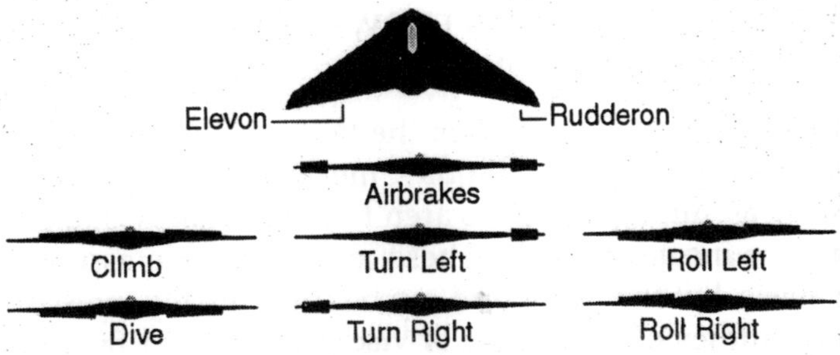

Flying Wing Control System

Incidentally, moving the rudderons away from their neutral position greatly increases the B-2's visibility to radar, so when it is flying into a combat area it steers by adjusting engine thrust, though this is more troublesome and less effective than using rudderons.

Aircraft may also have a number of auxiliary control surfaces. "Flaps" can be extended below the wings to increase the downwash and lift for take-off or low-speed flight, with flaps cascaded in some cases to increase their length. Control surfaces named "spoilers" or "lift dumpers" can be attached to the top of the wing to disrupt airflow and destroy lift, helping the aircraft to land. High-altitude aircraft have very wide wingspans and tend to be reluctant to come down on the runway, so their wings are usually fitted with lift dumpers.

Like a submarine, an aircraft has to maintain trim. This is obvious in the case of cargoes, which clearly can't be loaded onto a cargo plane in an unbalanced fashion, and the crew of a large cargo aircraft will generally include a specialist called a "loadmaster" who ensures that the cargoes are balanced properly incidentally, maritime cargo vessels also have loadmasters to reduce the hazard of capsizing due to an unbalanced load. Aircraft trim is less obvious in the important case of fuel management.

Usually aircraft designers will try to put fuel tanks at the centre of gravity of an aircraft, but an aircraft may have several fuel tanks, and care often must be taken to empty the tanks in a specific order, or the aircraft may become unbalanced. This was done manually by the pilot in the past, but now sophisticated aircraft have automatic systems to ensure the proper sequencing.

Since an aircraft will always have slight imbalances, instead of having the pilot try to compensate for trim faults by keeping the control surfaces slightly activated at all times, which would be very tiring, traditionally aircraft have featured small control surfaces called "trim tabs" that can be set to a fixed position using a dashboard knob to keep the machine flying level. Some aircraft may instead use knobs to introduce a slight bias in the aircraft's flight control surfaces.

COMPLICATIONS: VISCOUS FLOWS AND COMPRESSIBLE FLOWS

The discussion to this point on the flows of liquids and gases which can be lumped together as "fluids", meaning both can flow has been simplified by assuming the flows aren't viscous that is, subject to friction forces and are incompressible. These are reasonable assumptions for a simple analysis, but the real world can be substantially more complicated.

Formal studies of the effects of viscosity on fluid flow began in the middle of the 19th century, leading to the invention of equations for describing the viscous flow of an incompressible liquid; the equations were developed independently by Claude Louis Marie Navier, a French engineer, and Sir George Gabriel Stokes, a British mathematician, and so these rules are known as the "Navier-Stokes equations".

They're not simple equations, and in fact they are only fully solvable for simple scenarios. Bernoulli's principle no longer strictly applies since the viscous friction dissipates energy of the flow, resulting in a pressure drop along the pipe. At the outset, it was thought that this pressure drop would increase in direct proportion to the flow velocity that is, double the flow velocity and the pressure drop doubles as well but experiments showed that this was only true up to a certain threshold.

Beyond that threshold, the pressure drop increased with roughly the square of the flow velocity. This hinted that there were two different processes at work, which was demonstrated by Osborne Reynolds, a professor at the University of Manchester in Britain. Osborne devised an apparatus in which he could observe the flow of dye in water pumped through an tube.

He observed that the dye would flow straight for a distance, what was called "laminar" flow, and then break up into a tangle of eddies, what was called "turbulent" flow. This phenomenon can be observed by watching smoke float up off the tip of a lit cigarette: the smoke will rise for a space and then break into a tumble of coils. Incidentally, turbulent fluid flows are an example of a chaotic system.

In any case, the drag of the flow increases dramatically when the flow changes from laminar to turbulent. Reynolds was able to devise an equation that factored in the flow velocity, the density of the fluid, the viscosity of the fluid, and the length of fluid flow to return a value that would indicate when the flow changed from laminar to turbulent. The "Reynolds equation" is given as follows:

Fluid velocity × length of fluid flow × fluid density/fluid viscosity

The value returned by the Reynolds equation is known, to no great surprise, as the "Reynolds number (RN)". If the RN is less than 2,100, the flow will always be laminar. It doesn't necessarily switch to turbulent just above that value, but it becomes increasingly unstable and likely to shift to turbulent as the RN increases beyond 2,100.

One of the interesting features of the RN is that fluid flows with similar RNs behave in much the same way, no matter if the fluid is air or water or whatever. A sailboat moving at low speed in a dense fluid like water has much the same behaviour in terms of fluid dynamics as an aircraft moving much faster in a thin fluid like air. This has a major, though not obvious, implication for issues of scaling in aircraft design.

The smaller an aircraft gets, the greater the effective viscosity of the air. To a gnat, the air feels thick and viscous, and flying is more like swimming in fact, the smaller the creature, the less bother it is to get into the air, and small spiders will sometimes migrate by simply throwing out a thread of silk to the wind. Researchers working on tiny robot aircraft have finding out that conventional aircraft designs are not very workable at such small scale, and they have gravitated towards unusual designs, such as circular flying wings, or even "ornithopters" that flap their wings to fly.

The complexity of the analysis of viscous flows meant that scientists and engineers really needed to obtain simplified but still workable methods. In 1904, the German engineer Ludwig Prandtl observed that flows could be often be separated into two regions: a thin region known as the "boundary layer" close to the surface over which the flow passes and where the viscous effects are concentrated, and the region of flow above the boundary layer where viscous effects could be disregarded.

It is this viscous boundary layer that causes the airflow over the top of a wing to follow the curve of the wing in effect, it causes the airflow to stick to the wing. At low speeds, the boundary layer will tend to "separate" from the wing and ruin lift, and so a number of combat jets were built with a "blown flaps" or "boundary lsayer control (BLC)" scheme in which jet engine bleed air was blown over the top of the wing or the flaps to ensure that low-speed separation didn't occur.

The boundary layer can also be a nuisance for combat jet aircraft. The engines of such aircraft require a continuous large airflow to provide a high level of thrust, and if the engine intakes are close to the aircraft's fuselage they may swallow the viscous ("stagnant") boundary-layer air, choking off airflow and engine power even leading to "engine stall" that can "flame out" the engine or, in the worst case, damage it. Such aircraft often have a "splitter plate" between the front of each engine intake and the fuselage; these plates are set slightly off from the fuselage and prevent the boundary-layer air from being sucked into the inlet.

Assuming that a fluid is incompressible is appropriate if the fluid is a liquid, but it can lead to drastically wrong conclusions when the fluid is a gas. Even assuming nonviscous laminar flow, Bernoulli's law only really applies to slow-moving gas flows; if the gas flows are rapid, compression comes into play, leading to distinctly different behaviour.

Analysis of a compressible flow is too tricky to be discussed in detail in this document, but a simple example will illustrate the principles involved. Consider a standard rocket nozzle: it can be modeled as a cylinder that leads to a constriction, known as the "throat", which leads into a widening "exhaust bell" open at the end. Hot gases flowing down the hit the throat, where the velocity picks up and the pressure falls, roughly as it would in a fluid, though compression occurs as well.

Once the hot gases exit the throat and enter the widening exhaust bell, they expand rapidly, and this expansion drives the velocity up while the pressure continues to fall exactly the reverse scenario of what would occur if a fluid passed through a constriction into a widening bell.

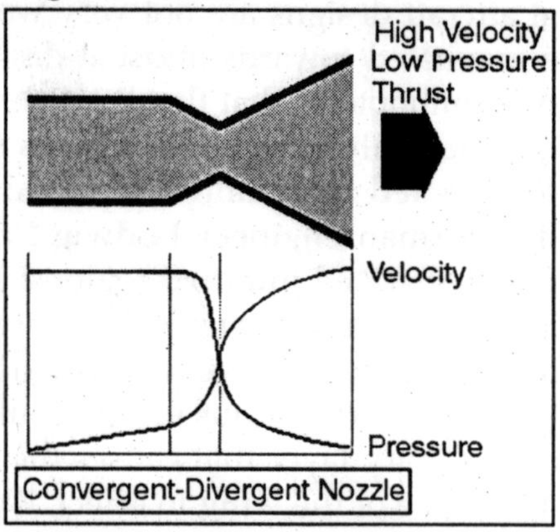

A standard rocket nozzle is designed to provide optimum thrust for a certain range of altitudes and atmospheric pressures. High-performance jet combat aircraft must operate effectively over a wide range of conditions, and so they have exhausts whose geometry can be varied depending on need.A fully "authentic" computer simulation of a real-life gaseous flow that takes into consideration viscous effects, turbulence, and compression can require some serious computing horsepower.There is a complicated interaction between all the facets of the simulation which can get even more complicated if chemical transformations take place in the stream.

One aspect to the complexity is that the pressure, volume, and temperature of a gas are closely related to each other. This leads to a consideration of concepts of "temperature" and "heat", which are somewhat devious to define.

Although this document has mentioned temperature effects in a casual fashion, to go much further requires a more detailed discussion. Similarly, although matter can often be converted between solid, liquid, and gaseous forms through changes in temperature, that also requires a more detailed look.

Chapter 3

Reality of Atoms

SPECULATIVE ATOMISM

Actual proof of the reality of atoms is an achievement of this century; even at its beginning critical consideration required the confession that atoms, although investigators spoke of them, were still no more than a stimulating and useful hypothesis, which in the final analysis was unproved and possibly unnecessary, perhaps even misleading. Although the idea of atomism was important and fruitful at the very beginning of western natural research, this idea did not originate from western research itself; just as we borrowed the model for mathematical thinking from the Greeks, we took the atomistic interpretation of nature from them.

Nothing exists — Democritus taught — except atoms and vacuum; everything else is imagination. These innumerable atoms, indestructible and invariable elementary constituents, are to be considered the basis for all beings and events in nature.

Individual atoms possess invariable geometric form and motions which fluctuate due to the pressure of and collision with other atoms. In all changes in the structure of nature the atoms are always preserved; from nothing comes nothing; nothing which exists can be destroyed; change is merely combination and separation of parts.

The differences of all things stem from the diversity of their atoms in number, magnitude, structure and arrangement. Since the movements of each individual atom are regulated according to natural law, nothing arbitrary can happen in all nature; nothing happens accidentally, but everything from a cause and of necessity.

Our rough senses can not recognize these amazingly fine elements in their true nature and form, but experience their effects only vaguely. "Only in the imagination does sweetness exist, in the imagination bitterness, in the imagination heat, cold, colors; in reality nothing exists but atoms and vacuum."

This is marked evidence of that "decoloration" of the world, the assurance that all direct sensory observation is an illusion and that a knowledge of the true status of things must lead us to a picture in which nothing remains but the geometrical form and mechanical motion of the atoms.

These idea forms of materialistic philosophy introduce the possibility of considering all events in nature as results of strict regularity; the gist of the basic concept of natural scientific thought is anticipated here in general and is thoroughly formed for the first time.

On the other hand we must not forget to recognize how radically this philosophy opposes religious representations of its time. For the Greeks every stream was a God; in the springs lived the nymphs, in the woods the horned Pan, and in caves and caverns resided demons.

This entire mythological picture of the world, which suspected the arbitrary and incalculable influence of demoniacal powers in every natural process, was pushed aside by the powerful conception of a picture of nature of stricter regularity. Epicurus, who renewed Democritus' teaching, and who in his thinking and conduct of life combined a rationalistic, vigorous attitude with a very cultivated, fine-spirited being, was not so brutal as to deny existence completely to the Gods.

He let them continue as blessed, immortal beings, occupied with themselves, and never infringing upon the workings of the world — these workings being developed according to the mechanical lawfulness of atomic motions in strict, definite sequence. In a great didactic poem *"De rerum natura"* the Roman Lucretius explained completely the ideas of atomistic regularity for all the fields of nature known at that time. These teachings were made known to western scholars in connection with the humanistic studies of the Renaissance; Gassendi, principally, introduced Greek philosophy into western science.

Let us consider the ideas which Newton, together with Gassendi, applying the divergent theories of Descartes, formed about the construction of matter. F. Dannemann explains them as follows in his historical work [1] on the natural sciences: "He considered it most plausible, that it (matter) consists of solid, impermeable, moving particles.

Since natural bodies, e.g., water, are invariable in their properties, the particles of which they consist must neither be able to be used up nor destroyed. The variation of material things is to be laid exclusively to the separations, combinations and movements of those invariable particles."

These are, as is obvious, exactly the concepts of materialistic philosophy, almost unchanged, just as Democritus stated them. But Newton was — outside of natural science — in no wise an adherent of materialistic theories. For him these ideas meant nothing but a clue to physical research; he did not see in them the support of radical, world-viewing results.

EVALUATION OF THE ATOMIC CONCEPT

Dalton first recognized and effectively utilized the stimulating force of the atomic concept and its ability to guide future physical research. To him we owe the use of the idea of atoms for understanding basic regularities of chemistry. Chemistry makes a sharp distinction between mixtures of different substances and chemical combinations.

A mixture — as, for example, a solution of sugar in water or a mixture of nitrogen and oxygen — can include arbitrary (within certain limits) relative proportions of the constituents. In a nitrogen-oxygen mixture the portion of oxygen can be varied continuously by addition or removal; correspondingly, the mixture exhibits properties somewhere between those of pure oxygen and of pure nitrogen, depending upon the relative concentration.

Also, the mixture can be separated into its constituents through relatively superficial means. In a chemical combination entirely new properties appear, which can be quite dissimilar to those of the constituents. The proportions of the components of a chemical compound are quite stable and cannot be varied continuously.

Thus exactly 16 grams of oxygen always combine with 2.016 grams of hydrogen to form water. Another chemical substance, hydrogen peroxide, can be formed from 2.016 grams of hydrogen and just 32 grams of oxygen.

It is striking that exactly twice as much oxygen is utilized in the second case as in the first; and on the basis of the atomic concept Dalton found a clear explanation for such facts, which occur similarly in all chemical combinations. His explanation was as follows: The "molecule" of water, i.e., the smallest possible particle of water, contains only one atom of oxygen; the molecule of hydrogen peroxide however contains two atoms of oxygen, while it contains just as many (two) atoms of hydrogen as the water molecule.

Here we arrive at the distinction between atoms and molecules; we designate the smallest particles of chemical compounds as molecules, which for their part consist of atoms of the chemical elements. According to Dalton's ways of thinking a comprehensive study of proportions by weight in chemical compounds rendered the determination of the relative atomic weights of all the elements possible. The "atomic" weight of oxygen is arbitrarily designated by the value 16, whereupon hydrogen acquires the atomic weight 1.008 and every other element, similarly, receives a definite atomic weight derived from chemical weight relationships.

The regularities which the gases exhibit, which in the case of sufficiently small densities assume an "ideal" form, proved very helpful for the clarification of these ideas; for example, in the sense that a gas mass contracts to one-half its original volume when, with constant temperature, the pressure impressed upon it is doubled.

Criteria existed for the hypothesis that not only in compounds but also in a gaseous element such as nitrogen the particles moving at random through space are not necessarily identical with atoms; in nitrogen, for example, each "molecule" of the gas consists of two atoms. In the chemical combination of gases (with constant pressure and temperature) there is a simple relation between proportions by weight and volume relations (proportions by volume) between the chemical compound and the still uncombined constituents. To fit these facts Avogadro introduced the explanation that equal volumes of all gases measured at constant pressure and temperature always contain the same number of molecules. Based on this "Avogadro's principle" a "molecular weight" can be determined for each gas independent of chemical weight relations.

The molecular weight of a chemically homogeneous gas is equal to the mass of 22.4 liters of this gas measured in grams at one atmosphere of pressure and a temperature of 0° C. Experience shows that values obtained in this way correspond with the chemically defined atomic weights.

In many elements, e.g., the metals, this molecular weight is exactly the same as the atomic weight; thus metallic vapors are "monatomic". In the aforementioned nitrogen, however, the molecular weight is exactly double the atomic weight. And in water it equals 18.016 ($2 \times 1.008 + 16$) in accordance with its above-described chemical composition. These ideas are further verified by simple relations of molecular weights such as, "diffusion velocities", o'r by the characteristic difference between the specific heats of "monatomic" and "polyatomic" gases. None of these, it must be emphasized, are proofs for the correctness of the atomic concept; but they are proofs for its usefulness. The atomic idea describes very plainly a large number of important and comprehensive regularities, thus fadlitating greatly our progress with these phenomena.

Crystallography furnishes evidence of further regularities which are certainly clearly explained by the atomic idea, though they too do not lead to proofs for the reality of the atom. In the crystalline state of matter the atomic concept suggests a regular grouping of the atoms or molecules, alongside of each other and arranged in layers.

This view induces important results; on this basis the variety of possible forms of crystals are shown mathematically to be limited very considerably. The completion of this mathematical proposition yields the wonderful result that according to the atomic concept there should be no more nor less than 32 different "crystal classes" which are defined by different properties of symmetry.

Actually this confirms precisely the experience of mineralogists; there really occur in nature all the crystal forms (crystal symmetries) which are compatible with the atom idea, but not one single one which contradicts it. The so-called "law cf rational indices" discovered by the mineralogists formed an important adjunct to this original consideration, which is related only to the symmetry properties of different crystal forms.

If we imagine the crystal as built up of a regular stratification of layers of atoms, then its outer boundary surfaces could never lie so that they partially cut through the atoms. The limits for the possible positions of the boundary surfaces given by this view are exactly equivalent to the requirements which the law of rational indices places on the position of the crystal surfaces.

Such experiences must necessarily encourage the more and more energetic pursuit of the atomistic notion — which intrinsically was so clear — even though tangible proof for the reality of atoms was still lacking. Actually the theoretical development of the atomic concept was furthered extensively by the most able physicists.

We have already mentioned ideal gases and indicated how the gaseous state of aggregation should be represented on the basis of the atomic theory; the molecules of a gas are completely separate from each other; each one moves along with high velocity until it meets another one, and then, in elastic rebound, their direction of motion and speed change.

The continuous impact of countless molecules on the walls of the enclosing vessel produces the pressure of the gas, perceptible to us with our rough tools. These ideas were elaborated thoroughly (Clausius, Maxwell) and our understanding of the gaseous state of aggregation was developed very clearly and completely. Whereby the proof resulted that Avogadro's principle, first only speculatively suspected, must necessarily be correct if the atomic hypothesis proves at all valid.

The discovery of the energy principle (Mayer, Joule) further strengthened confidence in atomism. It was recognized that mechanical energy, which disappears through friction or the like, reappears in the quantity of heat produced — in such a way that a definite amount of consumed mechanical energy corresponds to a definite number of calories of heat.

Conversely, according to the same conversion relation, in steam engines quantities of heat are transformed again into mechanical kinetic energy. This conversion of mechanical energy into heat and its converse can be considered quite consistent, in the sense of "macrophysics". Whereupon one must establish that energy, which has proved indestructible, can assume both the form of a quantity of heat and of mechanical energy. The "macrophysical", "phenomenologic" heat theory thus arrived at is entirely sufficient for answering all the questions encountered in connection with heat engines, and with heat conversion in chemical processes.

But further thought is suggested to give the process of the conversion of mechanical energy into heat a clear interpretation in the light of the atomic theory. In a gas, as was represented above, the concepts of heat and temperature are not applicable to the individual molecules, but only to the gas mass as a whole. In an individual molecule only the kinetic energy, i.e., its velocity of motion, or at best also its rotation, can be changed.

Thus, if energy is added to a gas mass, which appears macrophysically as added heat, it can only mean that the average energy of motion of the gas molecules has increased. The exact determination of the connection between the average energy of motion of the gas molecules and the temperature of the gas is a part of the theoretical investigations on the "kinetic theory of gases", which has already been discussed.

Similarly, in a solid body, a crystal perhaps, the temperature and heat energy contained within it are conceived of as mechanical energy of its molecules or atoms. Although in their regular arrangement in layers no one atom can leave its place, fine vibrating motions within the crystal — possibly similar to the fine vibration of a heavily traveled steel bridge — are still possible.

If these vibrations within the crystal are so small that they are not perceptible to us as mechanical motion, we do note the energy of these motions — as heat content which manifests itself in the temperature of the body. (If the atoms contained in each small piece of the crystal move by very small amounts in random directions, the whole crystal appears motionless to our macrophysical senses.)

Apropos of such thought processes was Boltzmann's atomistic interpretation of a remarkable regularity in heat phenomena. If we imagine that a planet's motion is interrupted and the planet then is induced to move in the opposite direction with the same velocity, it will retrace its entire elliptical path in the opposite direction after the reversal. This is an example of the fact that all purely mechanical motions are, as we say, reversible.

The upward motion of a stone, thrown from the earth, (in the ideal case) until its-point of reversal is an exact temporal mirror image of the subsequent downward motion. Similar reversibility — which can also be denoted as a physical "symmetry of the positive and negative time orientation" — exists in all purely electromagnetic properties.

But when we move a body over a table surface against frictional forces and heat is generated thereby, there is no reason to expect that in moving the body backwards along the same line the converse is true — that the heat energy will change back to energy of mechanical motion. Or take an electric current which flows through a wire and produces heat — when it flows through the wire in the opposite direction it again produces heat; the heat energy does not change back into electrical.

These are examples of irreversible processes. Another example of such a process is the mixing of two liquids; in most cases these intermix by themselves, but this process never runs backward of itself. A further example of importance to us is the following: if a gas filled vessel is introduced into an evacuated one and is opened there, the gas diffuses throughout the entire available volume. It is not impossible to restore the original conditions; the larger vessel can be evacuated again by pumping and the gas can be compressed into the smaller one.

But this does not constitute an exact reversal — a temporal reflected image — of the first process; and it can be considered that for the restoration of original conditions after any irreversible process a definite "compensation" must be included (Clausius).

In the steam engine, which undertakes to convert quantities of heat into mechanical energy, there does not occur a simple reversal of the process of the production of heat through friction, but each steam engine functions according to a scheme similar to the following: a certain amount of heat is removed from a heat container maintained at a high temperature.

A fraction of this amount is changed into mechanical work, while the remainder is conducted into a heat reservoir of lower temperature. As "compensation", therefore, for the conversion of heat into work there occurs the transfer of a quantity of heat from a hotter body to a colder one — i.e., therefore a process which nature could perform "irreversibly" itself, since in heat conduction the heat always flows from higher to lower temperature.

The occurrence of such "irreversible" processes among heat phenomena posed a very difficult problem for the "kinetic theory of heat" — tracing back the laws of heat to the mechanics of atoms.

For purely mechanical processes were shown to be always reversible, which makes it appear impossible to trace these irreversible heat phenomena back to mechanical properties.

The solution of this difficulty, achieved through Boltzmann's perspicacity, runs as follows: in an aggregation of many similar particles, such as are attributed to every physical system by the atomic theory, there is no point to pursuing the motion of each individual atom or molecule; it is only important to us to observe the average statistical behaviour of large numbers of atoms.

Thus statistical concepts are necessarily introduced into the consideration and we must determine which (defined in a rough statistical sense) events are to be considered as especially probable (thus as occurring very frequently) in comparison with other conceivable processes.

For our example ths signifies that when the small gas filled vessel is opened in the large empty one "it is extremely probable" that the gas will stream out and distribute itself uniformly in the large vessel.

Strictly speaking it cannot be definitely known that this will happen; indeed, to view the processes of motion in the gas mass with the same complete certainty achieved for planetary motions one must know quite exactly the position and velocity of each individual molecule at the beginning of the experiment.

The fact that, instead of this, we simply denoted the initial condition of the experiment statistically stipulates that we can not predict its further progress with complete certainty, but only with very great probability.

The reversibility of pure physical processes, which according to the atomistic conception is the basis for macrophysical heat phenomena, depends upon the existence of an exact reversal, an exact temporal reflected image for every process — whereby it can happen that the gas mass distributed in the abovementioned large vessel contracts into the smaller one, in an exact reversal of the normal process.

No one can guarantee that this is completely out of the question; but we can show that such an event is extremely improbable — mathematically its probability is expressed by an infinitesimally small number. Thus it becomes clear that despite the fundamental reversibility of atomistic elementary-processes, yet, on a large scale, in macrophysical events, practically irreversible phenomena take place.

Another instructive example of an irreversible process is the mixing of two gases or liquids with each other. According to the kinetic-atomic theory this problem is somewhat similar to the one wherein two types of balls — red and white ones for instance — are thoroughly shaken up in a sack.

If the red and white spheres are carefully separated initially, they will become completely mixed up after further agitation. If, then, we continue to shake, it is possible in principle that the original separation of red and white balls will be restored — but practically this possibility is of little moment since it would require shaking for astronomically long periods of time before this "spontaneous separation" could be expected with appreciable probability.

A quantitative measure for the irreversibility of a thermodynamic process (from a purely macrophysical standpoint) can be specified in the form of "entropy" — a quantity which always increases to a maximum in irreversible processes and does not decrease again (Clausius). According to Boltzmann this entropy can now be defined through the atomic concept as a measure of the disorder in an atomic aggregate.

THE LIMITS OF THE DIVISIBILITY OF MATTER

Everyday experience, which appears to show unlimited divisibility of matter, through simple refinements makes it certain that atoms must be extraordinarily small, and that therefore the number of atoms in a gram of any substance must be enormously large. For an exact measure of this quantity we use the so-called Loschmidt number.

As a "gram-atomic weight" of an element we denote a mass of a number of grams equal to the atomic weight; the Loschmidt number is the (equal for all elements) number of atoms in a gram-atomic weight.

Evidence for the divisibility of matter can be obtained by distributing a minute amount of a strong-scented substance (mercaptan) in a large quantity of air and then determining to what degree of dilution the odor can still be perceived; or by dissolving a strongly colored liquid (eosin) in a relatively large quantity of water and then noticing to what degree of dilution a uniform coloration of the water remains perceptible.

A still more suitable experiment involves the use of a very dilute solution of fluorescein, a substance which fluoresces strongly in incident light. Extremely small volumes of the solution are examined with a microscope to determine to what degree of dilution a spatially uniform fluorescence can be detected. From such observations it can be inferred (Perrin) that the Loschmidt number is definitely greater than 10^{21} (a 1 with 21 zeros after it).

But quite simple, daily experiences occur that indicate the limits which its atomistic structure places on the subdivision of matter. These involve utilization of very thin membranes of matter.

Gold leaf, which is used for gilding purposes, is hammered down to an exceptional thinness — when held against the light it exhibits a greenish translucence — of about 100 millimicrons (a million millimicrons make a millimeter); and yet this thickness is still far removed from the ultimate limit.

But there exist direct indications for the atomistic constitution of matter in commonplace, familiar soap bubbles, often formed when washing by a film of soapy water stretched between the thumb and index finger. These bubbles iridesce in variegated colors, whose diversity and rapid shifting is evidence of the variations in the thickness of the bubble with position and time. Small dark spots appear in a thus colored bubble, which at first may be considered as holes, but which in reality are enclosed by still thinner soap films.

If bubbles like this are produced in a solid frame instead of in the hand, and if they are protected from evaporation in a vapour-filled enclosure, they can be preserved for days. Newton had disclosed that inside these black spots — i.e., this membrane which weakly reflects light — still blacker, thus still thinner membranes are formed.

The thickness of the thinnest membrane obtained by Newton was about 6 millimicrons, about 20 times thinner than gold leaf. But the next thicker membrane obtainable, just before this thinnest one, is of exactly double thickness.

This provided a tangible indication of the molecular structure of this skin substance; obviously, the thinnest membrane represents a single layer of molecules, while in the next thinnest one two layers are piled together. A very thin film can be produced more conveniently by allowing a minute amount of oil to spread out on a large water surface.

The familiar iridescent films which oil forms on water correspond in their thickness approximately to the usual soap bubble. Much thinner oil films which are no longer directly visible but can be recognized with simple tools can easily be formed.

In these thicknesses of about 1 millimicron have been reached; and in these it has again been shown that the thicknesses of the finest films are not continuously variable. As in the soap bubbles, we find a quite definite thickness prescribed for the thinnest, second thinnest, etc. oil film (for a definite previously determined type of oil). The limits of the divisibility of matter have actually been reached here.

Further highly informative investigations were performed on very small corpuscular particles. Here the invention of the ultra-microscope rendered important service; it made visible not only the form, but also the position and motion of particles which cannot be "seen" in the usual sense, since there are limits, defined by the nature of light, for the optical observation of very small objects.

Bodies of dimensions smaller than the wavelength of visible light cannot be "formed" in any way by light rays; cannot be made visible in their true form. The ultra-microscope, through an ingenious device, made it possible to detect optically the presence of particles which are somewhat smaller than the wavelength of visible light.

For the fundamental task — to prove the reality of atoms experimentally — two types of investigations in particular have rendered real contributions. If very small particles are placed in a liquid it appears that they do not sink rectilinearly to the bottom and remain there but rather they move through the liquid in an irregular, erratic manner.

The motion becomes more intense the higher the temperature of the liquid is raised; but it is independent of a minor agitation of the enclosing vessel or of other accidental disturbances. Also, if left undisturbed, the erratic motion of the particles continues for days or years.

This Brownian movement serves as a direct proof for the correctness of the proposition, set up by the atomic theory, that the heat content of a body is to be conceived of as an internal motion which is so fine that it cannot be macrophysically detected as such. One must not conclude that there are very fine motions or currents in the liquid which produce the Brownian movement of the suspended particles.

That would lead to the idea that the motions of a throng of closely neighboring particles were somewhat similar as is the case with dust we see around us in the sun's rays. In Brownian movements two approaching particles in a liquid are completely independent of each other, and thus separate again very soon.

This shows that the hidden motions in the liquid, the permanent presence of which must be regarded as characteristic of the heat condition of the liquid, are completely "irregular" and macrophysically absolutely imperceptible.

Brownian movements have become the subject of many experimental investigations, for theoretical considerations had shown (Smoluchowski, Einstein) that its precise observation permits definite conclusions about the motions of atoms and molecules, through irregular collisions with which the particles in Brownian movements are driven around. It was possible to infer from investigation of Brownian movements how great the average kinetic energy of the individual molecules must be.

And since the total heat energy contained in the liquid is equal to the sum of the energies of the individual molecules, we can infer how many molecules there are in a given volume. This leads to a determination of Loschmidt's number.

Related investigations and considerations are also feasible in many kinds of "fluctuation phenomena". Imagine an armor plate suspended from chains and blown upon by a very strong, uniformly operating sand blast apparatus; the plate will be forced from its equilibrium position by the pressure of the blast and will then hang in a slightly different position.

Now, we remove the sand blast apparatus and by bombarding the plate with machine guns exert the same total pressure on the middle of the plate; the armor, struck over and over again by the single shots again assumes the displaced equilibrium position, this time continuously oscillating irregularly about this equilibrium position.

Observing just these oscillations, one can determine the strength of the single blows striking the plate. Similarly in many types of physical apparatus very fine irregular fluctuations can be observed and from them the magnitude of the Loschmidt number can be determined.

In the very fine physical apparatus the presence of the irregular heat motion of the atoms can be detected; thereby most varied methods can be utilized for determining Loschmidt's number. The values of Loschmidt's number obtained from all such investigations have always agreed within the limits of their accuracy.

The following describes the second possible way of determining Loschmidt's number from investigations on very small particles: the density of the earth's atmosphere decreases, as we know, with increasing height. Imagine two very high cylinders — about 100 kilometers high — to be filled respectively with two different gases, perhaps oxygen and nitrogen.

It is demonstrated that in the gas whose molecules are heavier (oxygen) the density decreases outward from the earth's surface more rapidly than in the other gas.

The gas molecules are hindered by their kinetic energy from simply accumulating on the bottom of the vessel, i.e., forming a liquid. But the heavier they are, the stronger is the effect of gravity forcing them downwards and thus the more rapidly does the gas density decrease with increasing height above the earth's surface.

The mathematical law relating to this states that the comparison of the decrease in density in the two cylinders depends solely on their molecular weights.

This was most useful in its application to the production of an artificial gas in which the actual mass of the individual molecules is known. Perrin laboriously produced large quantities of small resin pellets of equal weight. He obtained these resin particles from an emulsion which originally contained particles of the most varied magnitudes; but by an ingenious procedure, involving painstaking work, he was able to separate out particles of corresponding mass whose diameter was from 200 to 300 millimicrons.

These particles were then placed in water. (We know that with sufficient degree of dilution dissolved substances, e.g., sugar in water, behave analogously to the ideal gases.) Thus, the resin particles placed in the water represented an ideal gas and vessels a kilometre high were no longer needed to determine the decrease in density with increasing height with this artificial gas; even in small containers these artificial gas molecules, enormously heavy in comparison with usual molecules, crowd together noticeably at the bottom.

Through measurement of this effect and comparison with the known decrease of atmospheric density with increasing altitude it is possible to compare the directly determined mass of these artificial, huge molecules with the masses, absolute values still unknown, of the chemical molecules present in the atmosphere. Thus a new approach to the weighing of molecules — or differently expressed, to a determination of Loschmidt's number — is reached.

The numerical value which resulted from these different determinations of Loschmidt's number (and from further determinations still to be discussed) is 6×10^{23} (a 6 with 23 zeros after it). If the hydrogen atoms contained in 100 grams of water were distributed over the entire earth's surface, one atom would fall on each square centimeter.

Finally we come to the experiments in which single atoms are actually isloated and made, in a manner of speaking, tangible, so that there is no longer any possible doubt concerning their real existence.

Let us begin with Laue's famous discovery of crystal interference. Modern workers in physical optics and spectroscopy no longer use only prisms for the spectral resolution of light; the diffraction grating has become much more important.

When numerous lines at equal, close intervals are scribed upon a polished surface, a monochromatic (including just one wave length) light ray will neither be reflected from this surface as from a usual mirror nor as from a rough surface which reflects diffusely toward all sides.

The ray is reflected in a series of definite directions which vary according to its wave length; in all other directions there is no reflection because the light excitations traveling in these other directions mutually annul each other by interference.

This knowledge (a special interference experiment) gives us, as previously mentioned, the clearest and most obvious proof for the wave character of light and makes it possible to determine the wave length of the light just from, a knowledge of the spacing of the grating lines. (In common gratings this is between a hundredth and a thousandth of a millimeter.)

Conversely naturally the separation of the rulings on the grating, if by chance this value is not known, can be determined from the reflection produced by this grating with light of known wave length.

X-rays, on traversing crystals, exhibit remarkable interference effects which are related to those of the "line grating" described but are considerably more complicated. The discovery of these effects proved two things with one stroke; first, that X-rays are a wave radiation; and second, that crystals possess a fine grating-like structure, which is so fine that it defied observation with former tools.

Careful analysis of the varied and complicated interference phenomena which can be obtained in this way led to the certainty that this internal fine structure of crystals is exactly the same as that which atomistic concepts led us to expect; the "illumination" of crystals with X-rays clearly displays their synthesis from atoms and thereby conclusively assures the reality of these atoms.

The wave lengths of X-rays are much shorter than those of ordinary light (X-rays, like light, are also included within the general bounds of the classification of Maxwell-Hertz electromagnetic waves). Because of their short wave lengths it is not possible without further refinement to attain diffraction (interference) of X-rays with the usual, relatively rough optical line gratings.

But in crystals nature presented us with "natural diffraction gratings" with which we can study X-ray interference effects. With a very refined device it was possible to obtain diffraction of X-rays with an optical diffraction grating (line grating) so that the wave lengths of X-rays could be measured just as optical wave lengths are.

If we know the wave length of an X-ray, from the interference effects which result from the passage of this X-ray through a crystal we can infer the volume relations within the crystal. X-ray interference effects in crystals not only substantiate the reality of atoms with indubitable clearness and distinctness, but also facilitate another determination of the number of atoms in a macroscopic piece of matter — a new determination of Loschmidt's number.

Millikan (in extending and refining older, less lucid ones) performed an experiment which was striking in the simplicity of its fundamental idea. In it he provided tangible proof for the atomic nature of electricity. A mist of minute oil drop lets is sprayed into a chamber; many of these are electrically charged in the process of their production.

One droplet is singled out and observed through a microscope. Left to itself it falls vertically downward, due to the force of gravity; not according to the ideal laws of free fall, but rather with constant velocity, since air resistance is proportionately more effective against these small droplets than against larger bodies.

If the droplet is subjected to the influence of an electric field, of proper intensity perpendicular to its direction of fall and opposing it, the droplet will start to rise, with a constant velocity, the magnitude of which is dependent upon the field strength.

The ratio of the forces working in both cases and the magnitude of the charge carried by the droplet can be calculated from a comparison of the speeds of rising and falling. Consideration of the values obtained for numerous such droplets indicates that the magnitude of the charge does not vary continuously from case to case; none of them ever has a smaller charge than the so-called "elementary charge", and larger values are integral multiples (2, 3, 4...) of this smallest one.

Thus we have tangible evidence that electric charges cannot be artibitrarily divided indefinitely; rather, all electric charges are composed of indivisible "elementary charges", each of which carries a charge of 4.77×10^{-10} (4.77 divided by a 1 with 10 zeros after it) electrostatic units.

This atomism of electricity is closely related to the atomism of matter in general. A law concerning electrolysis, discovered by Faraday and named after him, states this relation. Helmholtz had early concluded from this law that if the concept of the atomism of matter were proven correct an atomism of electricty must be assumed.

If one zinc and one copper plate are immersed in an aqueous solution of copper sulfate and an electric current is passed through the solution in a positive direction from the zinc to the copper plate, the zinc gradually becomes dissolved in the liquid as further copper is deposited on the copper plate.

Quantitatively this experiment shows that when one gram atomic weight of zinc has been dissolved, just one gram atomic weight of copper has been deposited; and corresponding weight relations are found to hold for all similar electrolytic experiments. Moreover, the number of gram atomic weights deposited or dissolved always remains in a simple proportion to the quantity of electric charge which has passed.

Thus there occur here regularities of the same character as the laws of weight relations in chemical reactions which Dalton explained by the concept of atomism. In these laws Dalton saw a criterion for the definition and use of the atomic concept; by the same right we can, with Helmholtz, infer the atomistic structure of electricity on the basis of Faraday's law.

Conversely, knowing from Millikan's work that the atomism of electricity not only expresses an auxiliary idea but is a real fact, we can infer from Faraday's law that the atomic structure of matter is a reality (which has already been verified in our consideration of crystal interference).

The knowledge of the magnitude of the elementary electric charges furnishes one more basis for calculation of Loschmidt's number, characteristic of this atomistic structure of matter.

This in no way exhausts the experiments which show the indubitable reality of atoms. Further direct proofs for atomism are gained in connection with the investigation of radioactivity.

From the observation of radioactive preparations it was determined that "alpha-radiation" is simply an emission of electrically charged helium. When this alpha-radiation strikes a suitably prepared zinc sulfide screen, even with limited intensity, it produces a series of small light flashes (scintillations) there.

From this effect it is seen that the emitted electrically charged helium is not an arbitrarily divisible, indefinitely diffuse substance, but must consist of discrete particles. By counting the light flashes, it became possible to determine how many single atoms constitute a definite quantity of helium; a new method of determining Loschmidt's number. The results again confirmed those obtained by other methods; this confirmation is evidence that each of these small light flashes actually does come from a single (electrically charged) helium atom.

Of even greater significance was still another process, in which again the effects of single atoms (or "ions", as electrically charged atoms are usually referred to) became visible. A moisture saturated atmosphere is produced in a carefully cleaned, dust-free vessel.

If the size of the container is suddenly increased by means of a moving piston the saturated atmosphere cools off and becomem "super saturated" — the water vapour begins to condense to liquid droplets.

But the first droplets require "condensation nuclei" on which to form, and by removing all very fine particles (dust, etc.) we have eliminated the usual nuclei.

If an alpharadiation from a radioactive preparation strikes the atmosphere of this "Wilson Cloud Chamber", each of the rapidly traveling helium ions produces a fine track, along which are present innumerable atoms or molecules, now likewise electrically charged, which were hit by the alpha particles shooting through the air.

Along the entire path these newly produced ions form suitable condensation nuclei; the entire path appears marked as a fine streak of mist visible to the naked eye.

Again the effect of a single charged atom — and this time still more beautifully and definitely than by scintillations -has become visible, and again calculations from such "cloud tracks" yield Loschmidt's number.

The Wilson chamber has become one of the most important research tools of modern atomic physicists. Obviously its usefulness is limited to very rapidly flying particles, since only these possess sufficient energy to "ionise" innumerable other atoms or molecules along long paths.

But in the processes connected with radioactivity these particles appear under various circumstances; and the Wilson chamber presents a beautiful opportunity to study all the details of the processes associated with their formation and transformation.

Geiger's development, the counting tube, is an equally important tool for atomic physics — for determining effects of single, isolated atomic particles. As soon as a single rapidly moving electrically charged atom enters this apparatus a macrophysically perceptible electric charge is produced by an ingenious arrangement.

Like the Wilson chamber this apparatus makes it possible to establish the reality of atoms, to determine Loschmidt's number and to study thoroughly atomic physical processes.

CONTRIBUTIONS OF ATOMIC PHYSICS

Before Millikan's determination of the elementary electric charge, electricity — notably through investigations by Lenard had been represented in pure culture, so to speak, namely in the form of cathode rays formed in a highly evacuated electric discharge tube — also in an X-ray tube, for example, where they produce X-rays by impinging on a solid plate.

These cathode rays consist of a flow of negative electricity; from the results of Millikan's experiment it is certain that this electricity must flow in the form of separate, indivisibly small particles. These particles have been named "electrons"; they might also be designated as the atoms of electricity.

The cathode ray which travels rectilinearly when undisturbed can be forced into a curved path through the use of electric and magnetic fields. The motion of an electron deflected under the influence of known electric or magnetic fields must be calculable by Newton's second law — force equals mass times acceleration — if we know the ratio of its charge to its mass, since the acting force is proportional to the charge on a single electron.

Conversely, this relation of charge to mass can be inferred from the experimental determination of the deflection of cathode rays in electric and magnetic fields.

Since the charge on an electron is known from Millikan's experiment (naturally from the outset it is assumed highly probably that the electrons in a cathode ray possess a single Millikan elementary charge, not two or three; further experiments confirm this as fact) the mass of an electron can be calculated from the measurements on cathode rays.

The result shows that the mass of an electron is about 1838 times smaller than that of a hydrogen atom; or, expressed in grams, equals 0.9×10^{-27} (0.9 divided by a 1 with 27 zeros after it) grams.

The method of determining the ratio of charge to mass with cathode rays can also be extended to electrically charged atoms, ions. We know that from hydrogen atoms only one type of ion can be formed — a particle which has almost the same mass as the hydrogen atom itself and possesses exactly one positive elementary charge.

Two different forms of ions — each of almost the same mass as the atom — can be formed from the helium atom; one has a single positive elementary charge, the other has two. The impression results that the electrically neutral hydrogen atom contains positive and negative charges within its structure. Obviously hydrogen contains just a single positive and a single negative electric charge; and the negative one must be connected with a proportionally minute mass, whereas almost the entire mass of the conversely, this relation of charge to mass can be inferred from the experimental determination of the deflection of cathode rays in electric and magnetic fields.

Since the charge on an electron is known from Millikan's experiment (naturally from the outset it is assumed highly probably that the electrons in a cathode ray possess a single Millikan elementary charge, not two or three; further experiments confirm this as fact) the mass of an electron can be calculated from the measurements on cathode rays.

The result shows that the mass of an electron is about 1838 times smaller than that of a hydrogen atom; or, expressed in grams, equals 0.9 x 10^{-27} (0.9 divided by a 1 with 27 zeros after it) grams.

The method of determining the ratio of charge to mass with cathode rays can also be extended to electrically charged atoms, ions. We know that from hydrogen atoms only one type of ion can be formed — a particle which has almost the same mass as the hydrogen atom itself and possesses exactly one positive elementary charge.

Two different forms of ions — each of almost the same mass as the atom — can be formed from the helium atom; one has a single positive elementary charge, the other has two. The impression results that the electrically neutral hydrogen atom contains positive and negative charges within its structure.

Obviously hydrogen contains just a single positive and a single negative electric charge; and the negative one must be connected with a proportionally minute mass, whereas almost the entire mass of the hydrogen atom is attached to the positive one. The supposition is suggested — and abundant experience raises it to certainty — that the negative charge in the hydrogen atom is just a negative electron.

The positive hydrogen ion, from which no further electrons can be separated, is designated as the "nucleus" of the hydrogen atom; it is also referred to as a "proton". In the helium atom two negative electrons are needed to neutralize the positive charge of the "nucleus"; here again ionization of the atom means simply the removal of one or both electrons. The nucleus of the helium atom is, moreover, identical with the aforementioned alpha particles.

As Rutherford recognized, the structure of all atoms is similar to these. In each atom almost the entire mass is concentrated in a positively charged nucleus; the effect of the "envelope" or "cloud" of negative electrons surrounding this nucleus is to render the atom electrically neutral. Thus the number of these electrons must always be equal to the "nuclear charge number"; i.e., equal to the number of positive elementary charges in the heavy nucleus.

Chemists have known for a long time that a very convenient view of the chemical elements can be obtained through their arrangement in the "periodic table of the elements" -wherein the elements are arranged (Meyer and Mendeleef) according to their atomic weights.

It was later shown that the atomic weight is not the characteristic by which the various chemical elements are differentiated that atoms of different atomic weights can belong to the same element (isotopes) and that atoms of equal atomic weights can belong to different elements (isobars).

Not the mass of the nucleus, but its charge really determines the chemical nature of an atom; for the atomic number (nuclear charge number) determines the structure of the electron shells since it dictates how many electrons the neutral atom must possess. Through the structure of the electron shells the chemical properties are determined. Chemical molecules are formed when two or more atoms combine; in the combination the outer rings of electron shells experience certain transformations, but the nuclei remain unchanged.

The size of an atom can be determined in many ways. Their diameters are all equal to about 0.1 millicron. These diameters cannot be specified by exact numbers like the masses can, since the atoms are certainly not smooth spheres with definitely defined surfaces. As in a cloud with blurred boundaries, in an atom only an approximate value can be specified for its diameter. Such values can be determined from crystals, in which the atoms must be packed together very closely.

From investigations of gases it is possible to obtain corresponding determinations of these atomic magnitudes. The atomic nuclei are very much smaller; perhaps a hundred thousand times smaller in diameter than the atoms. Thus, almost the entire volume of an atom is occupied by the electron shells.

But an electron is essentially no larger than an atomic nucleus. How it is possible that despite this, in a hydrogen atom for example where the entire "shell" consists of only one electron, the volume of the entire atom is "filled" up by this electron is a question to which we must return.

While chemical processes involve only the outer electron shells, there are other processes in which the nuclei themselves experience transformation, division or synthesis. In the main these are the already noted processes of radioactivity; but also numerous artificial, arbitrarily produced transmutations have been produced in atomic nuclei in recent years.

Thus the nuclei must also be represented as complex structures; today we trace all nuclei back to two building stones, one of which is the already described proton and the other the "neutron" — a particle which has nearly the same mass as a proton but is electrically neutral.

Chapter 4

Model of Atoms

RUTHERFORD'S MODEL

The perception of an atom as the least indivisible particle of substance was the foundation of the cinematic gas theory. It have explained and predicted a series of the experimental phenomena.

However, at the end of XIX century new experimental data proved the composite structure of an atom. The experimentally found laws of radiation and uptake of energy by atoms (spectral series, radiation of an ideal black body, photoeffect) have not received an explanation within the framework of a classical electrodynamics. Only new ideas, based on the quantum theory, helped explaining these phenomenon.

Structure of Atom

British physicist Ernest Rutherford (1871-1937) has created a planetary model of atom exploring a dispersion of a-particles transiting through a thin metal foil. According to this model an atom looks like a tiny planetary system, in which the forces of an electrical attraction operate.

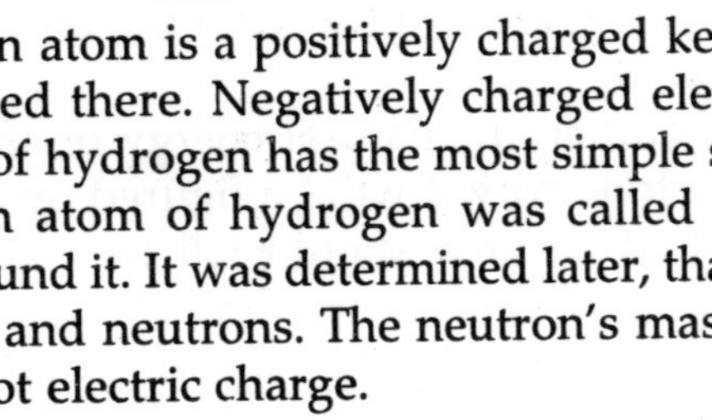

The centre of an atom is a positively charged kern. Almost all mass of an atom is focused there. Negatively charged eletrons rotate around the kern. An atom of hydrogen has the most simple structure. Positively charged kern of an atom of hydrogen was called a proton. The only electron rotates around it. It was determined later, that every atom's kern consists of protons and neutrons. The neutron's mass is almost equal to proton's, but has not electric charge.

The mass of an electron is about 1/1840 of a proton's mass and does not influence an atomic weight of substance. The number of positive and negative charges of atom determines the position of a substance in a periodic system. The positive charge of an atom's kern is equal to a serial number of a substance in the table.

The number of neutrons in an atom's kern is equal to a residual between a rounded atomic weight and a number of positive protons in a kern (serial number of substance). The electrical neutrality rule of atom says, that the number of electrons rotating around the kern is equal to a number of positive protons contained in an atom.

Parameters of Atom

According to the Rutherford's experimental results the radius of a trajectory of an electron in a hydrogen atom is 0,53 × 10^{-10} m, the diameter of an atom of hydrogen is equal to 2,6 × 10^{-15} m. Volume of electrons and kern is approximately 1/10 000 of an atom's volume.

The substance of a kern has huge density of 200 million ton in 1 cm^3. It is all because the most of an atom's mass is concentrated in its kern.

The mass of a positive proton is approximately equal to the mass of a neutron and is 1,67 × 10^{-27}. The mass of an electron is 9,11 × 10^{-31} kg.

The classic radius of an electron, r_0, is equal to 2,8 × 10^{-17} m. It is the distance, on which the electrostatic energy of an electron intercourse with a point charge of the same quantity is equal to natural energy of an electron:

$$e^2 r_0 = m_0 c^2$$

e ⁻elementary charge; c ⁻speed of light in vacuum;

m_0 and nbsp- electronic mass

Deficiencies of the Theory

The Rutherford model of atom was well correlated with the results of a dispersion of an a-particle by atoms of substance, but it has not explained neither the process of radiation of atoms, nor legitimacies in spectrums of radiation. According to the laws of an electrodynamics rotating around the kern the electron should radiate electromagnetic waves.

As a result of radiation the natural energy of an electron should be diminished, thus the trajectory it will be figured by a spiral, and during the order 10^{-8} about an electron should fall on a kern. Such deduction obtained on the basis of representations of classical physics about radiation, contradicted known stability of atoms and character of nuclear spectrums of radiation.

BOHR'S MODEL

Structure of Atom

The Danish physicist Niels Henrik David Bohr (1885-1962) has created the theory of a structure of atom, which was deprived of the basic deficiencies of Rutherford's model of atom.

Thus he had to change some habitual representations of classical physics. Bohr has formulated three postulates (sometimes the first and the second postulates are combined, then Bohr's theory of atom is reduced to two postulates):

- The electrons in atom rotate around the kern on particular stationary orbits only. The motions on these orbits correspond to stationary states of atom, which do not vary in time without exterior actions. Rotating on stationary orbits, the electrons do not radiate energy.
- Allowed, discrete values of energy of an electron correspond to stationary states of atom.

This postulate was called the quantization rule of orbits. It affirms, that when the electron is going on a stationary orbit, it

Should have discrete values of a moment of motion:

$$m_0\, ur = n\frac{h}{2\pi}, \; n = 1, 2, 3...$$

$m0$ and nbsp- electronic mass u - velocity of atom;

r- radius of an orbit h- Planck constant;

$h = 6.62 \times 10^{-34}$

- The radiation or uptake of energy by atom occurs only at transition of an electron from one stationary orbit to another. Thus quantum of light energy (photon) is radiated or is immersed. The energy of a quantum is equal to a difference of energies of an electron on the relevant stationary orbits: frequency of a light, energy of an electron on m and n orbits.

The laws of classical physics say, that the transition of bodies from one state to another occurs continuously at continuous radiation or uptake of energy. The classical laws have appeared inapplicable to the phenomena, which are described by Bohr.

Parameters of Atom

The radiuses of stationary orbits and energies of electrons are determined by dependence of energy of atom on radius of an electronic orbit. The elementary system is the hydrogen-like atom. The hydrogen-like atom is the one, in which one electron rotates on a circular orbit around of a kern, which has a charge *Ze* (Z – serial number of a device in periodic system).

The energy of an electron in atom consists of a kinetic energy of a motion on an orbit and potential energy in an electric field of a kern (electronic mass, travelling speed of an electron on an orbit, radius of an orbit, permittivity of vacuum of empty space):

$$E = \frac{m_0 u^2}{2} + \frac{-Ze^2}{4\pi\varepsilon_0 r} = \frac{m_0 u^2}{2} - \frac{-Ze^2}{4\pi\varepsilon_0 r}$$

m_0 and nbsp- electronic mass; u - velocity of atorn

ε_0 -permittivity of vacuum of empty space

r- radius of an orbit

The kinetic energy is determined with the requirement, that a centripetal acceleration is created by force of a Coulomb attraction of charges. The kinetic energy of an electron is inversely proportional to radius of an orbit.

A total energy:

$$E = \frac{1}{2}\frac{Ze^2}{4\pi\varepsilon_0 r} - \frac{-Ze^2}{4\pi\varepsilon_0 r} = -\frac{1}{2}\frac{Ze^2}{4\pi\varepsilon_0 r}$$

The total energy of an electron varies with the change of a radius of an orbit the same as potential energy does. Hence, the more than radius of an orbit of an electron, which rotates around the kern, the larger is the total energy of atom.

Value r_1 = 0,528 × 10^{-10} m is called the first Bohr's radius. For the subsequent values: $r_n = n^2 r_1$

n - main quantum number. It specifies an energy level, on which the atom is. The lowest state of atom is inconvertible. All atoms are in the lowest state at standard conditions. The atom can be exited, if the energy transferred to it. Thus it will be transferred to one of higher energy states. The distance from the lowest energy level $n = 1$ up to the proximate level with higher energy n = 2 makes 10,1 eV = 16,2 × 10^{-19} J for the atom of hydrogen. This energy is the least bundle of energy, which the atom of hydrogen can immerse, when it is in a normal state.

Deficiencies of the Theory

Bohr's theory explained the nature of a spectral series of hydrogen. It has allowed to spot theoretically the sizes of atom and energy levels of electrons for atom of hydrogen. Deficiency of Bohr's theory is the inconsistency of its build-up. Bohr's postulates have quantum character, but the stationary electronic orbits and allowed levels of energy of atom are determined by methods of a classical mechanics and electrodynamics. Therefore Bohr's theory has appeared usable to calculations only of monovalent atoms.

Bohr's theory does not give an opportunity to calculate intensity of spectral lines, which is interlinked to probabilities of electronic transitions. The introduction of quantum rules for definition of orbits and levels of electrons was not logically justified by Bohr.

The further development of the Bohr's theory was made by German physicist Arnold Johannes Wilhelm Sommerfeld (1868-1951). He has changed circular orbits on elliptic.

It has enabled to apply Bohr's theory to multi electronic atoms. However the creation of a fundamental theory of atom is bound to the creation of a quantum mechanics.

MULTIELECTRONIC ATOMS

Structure of Atom

Just as in the atom of hydrogen, in more composite atoms electrons can rotate around of a kern on allowed orbits only. The orbits in composite atoms cluster in system of shells. Each shell contains a particular number of orbits, an electron can be on any og them. A *K*-shell is the closest to a kern and contains two orbits. The second is a *L*-shell has eight orbits. The third is a *M*-shell has eight orbits, the fourth *N*-shell has 18 orbits etc.

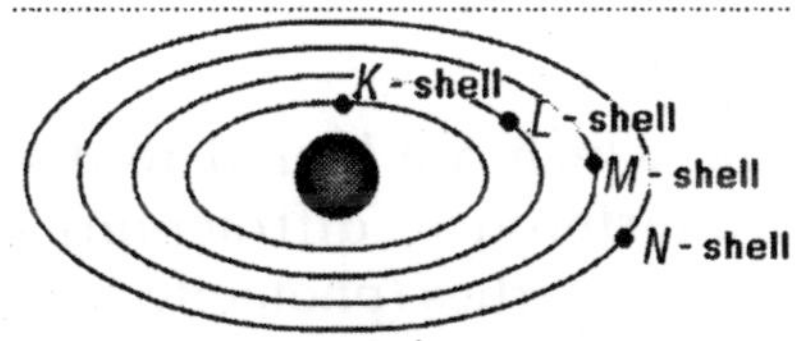

Parameters of Atom

When the electron travels from an orbit of greater radius to an orbit of smaller radius, the energy is oozed. But the transition from outer shells on interior is possible only when there are free orbits on the leter. The outer-shell electrons are less bound to a kern than interior ones: they are on greater distance from a kern, and the electrons, which are on inner shells, screen the nuclear charge.

Therefore just a small amount of energy is sufficient for excitation of outer-shell electrons. The quantums of energy radiated at transition of such electrons in a nonexcited state have a quantity of about several electron-volt. Greater energy is necessary to detach inner-shell electrons from the atom. The greater the nuclear charge the greater energy is needed. Therefore, when there is a transition of electrons from outer shells to a *K*-shell in heavy atoms, the emission of quantums with energy of hundreds and even thousand electron-volt is possible. Such quantums of high energy are called Roentgen rays.

X-ray Spectrums of Atoms

The X-rays results from losses by atom of an inner-shell electron: The transition of electrons is accompanied by an emission of quantums of high energy with a small wave length, which are termed as Roentgen rays.

The X-rays emerges when atom losses inner-shell electron: The transition of electrons is accompanied by an emission of quantums of high energy with a small wave length called Roentgen rays. The atom can lose inner-shell electrons as a result of an electron bombardment with electrons speeded up to energy of several thousand electron-volt. The wave length of the X-rays is comparable to interatomic distance in solid bodies.

The examination of the X-ray diffraction on crystal lattices of solid bodies enables, knowing a wave length of Roentgen rays, to determine interatomic distances in them. The examination of Roentgen rays also allows to determine nuclear charge, because it depends on a prime relation of an electron-binding energy in atom, which is expressed through energy of the X-rays. This relation is called Moseley's law:

$v = A(Z - B)$

v and -frequency of a line in a X-ray charateristic spectrum of a device

A and B - constants Z -serial number of a device

PROPERTIES OF LIGHT

In one phenomena, which are bound to distribution of light, its undular properties (interference, diffraction) are shown. In other phenomena its quantum properties (photoeffect, radiation and uptake of a light, process of interaction of a light with substance) are shown.

The light has simultaneously undular and quantum properties, which are differently shown in the different phenomena. However, when the wave length of a light is diminished, its undular properties are attenuated and the quantum properties are shown.

Compton's Effect

The experience of Arthur Holly Compton on a dispersion of Roentgen rays by substances, which consist of atoms of light devices, have shown, that the dispelled Roentgen rays have a major wave length λ', than impinging λ. Thus the residual $\Delta\lambda = \lambda' - \lambda$ depends on properties of scattering substance and wave length of an impinging light:

$$\Delta\lambda = \lambda' - \lambda = 2\lambda_K \sin^2 \frac{\theta}{2}$$

The quantity λ_K is a stationary value for all substances also is termed as Compton's wave length, $\lambda_K = 2{,}43 \times 10^{12}$ m. This phenomenon, which is termed as Compton's recoil, is impossible to explain by a wave theory of light. In quantum theory the magnification of wave length of radiation at a dispersion is explained by interaction of photons of impinging radiation with electrons of atoms and molecules.

The part of energy of an impinging photon, which is transmitted to electrons of scattering substance, depends on a scattering angle of photon.

At $\theta = \pi$, that is provided that after a dispersion a photon will fly in the party, which is opposite to a tentative direction, the losses by a photon of energy will be maximal, that will give in maximal magnification of a wave length of radiation.

Light Pressure

Some properties, for example light pressure, can be surveyed simultaneously in the electromagnetic theory and in the quantum theory. According to the electromagnetic theory of light, the pressure p, which renders a luminous flux with power W, impinging normally on a simple surface, is equal:

$$p = \frac{W}{c}(1+R)$$

For an ideal black body ($R = 0$): $p = \dfrac{W}{c}$

For an absolute reflecting bodyh ($R = 1$) $p = \dfrac{2W}{c}$

If N of quantums of a luminous flux with frequency ν, which bears(carries) energy $W = N \cdot h\nu$, impinges on a simple surface for 1 second, then the blanket impulse at a reflectivity R will look like:

$$p = \frac{N \cdot h\nu}{c}(1+R) = \frac{W}{c}(1+R)$$

Thus, electromagnetic theory and quantum theory give identical results. However it is essential different representations will be utillized in the electromagnetic and quantum theories of light. The concept of an electromagnetic wave of application to a light guesses existence of a wave in particular field(area) of space, which is capable to contain even some lengths of waves of light.

The light quantums are considered as a particle, which are localized in very small volume of space. The representation about a light wave guesses a continuous distribution of energy in a spatial wave, the representation about quantums guesses localization of energy in a small part of space.

Both these incompatible representations reflect unity of a wave-corpuscle nature of light, which is shown at distribution of light and its interaction with substance.

PROPERTIES OF SUBSTANCE

De Broglie's Theory

The French physicist Louis Victor duc de Broglie has applied wave-corpuscle treatment and to particles of substance: as the light detects corpuscular properties, also particles of substance should detect undular properties.

De Broglie has utillized already wave-corpuscle properties, accepted for a light, and their quantum rule and has expressed a wave length through the performances of a particle. For a light the corpuscular point of view determines energy and impulse accordingly by expressions $E = mc^2$ and $p = mc$.

The wave length of a particle of substance is determined under the formula:

$$\lambda = \frac{h}{p} = \frac{h}{m\upsilon}$$

At adding in a rule of quantization of a moment of momentum of an electron the impulse of an electron, will turn out a relation $n\lambda = 2\pi r$. From it follows, that on length of a stationary orbit of an electron the integer of lengths of waves should be stacked.

Thus, the requirement of existence of allowed orbits, which was injected by Bohr frequency relation, acquires legible physical sense: the allowed orbits are orbits supposing formation on them of standing waves of an electron. For a rectilinear motion of particles de Broglie has offered expression, which is made by analogy with the equation for a spread flat light wave in the complex shape:

$$\psi = \psi_0 e\ i(wt - \frac{x}{\lambda})$$

De Broglie's Waves

The specified waves have received a title of associated waves. Thus, the associated waves are traveling waves for freely propellent electrons or standing waves for electrons, which are bound in atoms. De Broglie's waves are not waves of propellent substance, they have not analog in classical physics.

De Broglie's guess about existence of undular properties of particles carries universal character: undular properties should have the electron, positive proton, neutron, atom, molecule, any moving object.

However objects, which have major masses and move with usual velocities, will have a so small wave length To in comparison with the sizes of objects, that the phenomena of an interference and diffraction for them can completely be neglected. The undular properties of a light are clearly shown in cases, when the wave length has compared to the sizes of bodies, with which the light interreacts.

The lengths of waves of electrons in usual requirements have the order of the nuclear sizes, hence, the effects are characteristic for them which are observed usually for Roentgen rays. For such small lengths of waves it is possible to observe a diffraction on nuclear crystal lattices.

EXAMINATION OF PROPERTIES

The existence of undular properties of substance was revealed at observation of patterns of a diffraction of beams of particles. Legible patterns of an interference and the diffractions give bundles of electrons, which have almost identical impulses. These patterns are very similar to optical patterns of an interference and diffraction. At change of quantity of a momentum of electrons and measuring of a wave length, it is possible to receive a relation between these quantities. It has appeared equal to a known relation.

$$0\lambda = \frac{h}{p}$$

This de Broglie's relation is valid for any kind of particles. The undular properties of particles were detected at observation of a pattern of the diffraction of the bundle of electrons on chips. At measuring a scattering angle it is possible to spot a wave length of electrons. Thus, at known quantity of potential, which was utilised for acceleration of a bundle of electrons, their velocity is possible to calculate, and then to receive independent value of a wave length of an electron. The value of a wave length of an electron, which was spotted in such a way, completely coincides with quantity of a wave length, which is found from a diffraction pattern.

The undular properties of electrons also were detected in experience with a cathode rays at major velocity, which was passed through a thin metal foil on a photographic plate. After development of a photoplate on it the diffraction pattern from rings of different intensity was visible. However this diffraction pattern was stipulated not by Roentgen rays, which could arise at a concussion of high-velocity electrons with metal. It confirms presence of a stream of charged particles.

MECHANICS

Quantum Mechanics is the study of matter and energy on the subatomic scale. It was came into being because classical physics could not explain certain experimental results, such as why the photoelectric effect occured only when the light shone is above a certain frequency. Quantum Mechanics and Relativity became the foundation of modern physics.

Main Points

Principle of Quantatization

The fundame ntal principle of Quantum Mechanics, and its namesake, is that energy came in discrete particles called "quanta". This idea, the Quantatization of Energy, was put forth by Planck in his Quantum Hypothesis(1900), an explanation of blackbody radiation. It was a departure from Classical Physics, which asserted that energy could be infinately small.

Photoelectric Effect

This quantatization of energy approach was used by Einstein in his 1905 explanation of the photoelectric effect, the emission of electrons by a metal when light is shone on it.

No electrons are emitted when the light is below a threshold frequency, no matter how intense the light. Einstein postulated that light, while exibiting wave properties as shone by the Double-slit Experiment, also came in the from of particles called "photons", discrete packets of energy.

The energy of a photon is proportional to the frequency of the light(E=hv); thus, light with higher frequency had more energy and could excite the electrons, while light whose frequency were below the threshold could not, no matter how many photons there were.

Bohr's Model and Spectral Lines

Bohr also used the Quantatization of energy in his 1913 model of the atom. He stated that electrons revolved around the nucleus much like the planets revolve around the sun. They could be found in predefined, stationary orbits, radiating no energy as they revolve.

Only when they gain a quatum of energy equivalent to the difference between the energy of the orbits do they become excited and enter a higher energy orbit. This is consistent with Einstein's explanation of the photoelectric effect, and explains why spectral lines only occur at certain places.

Wave-Particle Duality of Matter

In 1923 De Broglie expanded the Principle of Wave-particle Duality of Light put forth by Einstein to include all matter, called Principle of Wave-particle Duality of Matter. He asserted that all matter, not just light, exibit both wave-like and particle-like behaviour,

Quantum Matrix and Wave Mechanics

By now there was enough theories to put together a quantum picture of the atom. However, it was very difficult to sort through all the math.

Heisenherg and Shroedinger simultaneously and independently worked to put forth a complete quantum mechanics. Heisenberg employed use of matrices, and his model was called the matrix mechanics.

His approach emphasized the quantum—discrete properties while Schroedinger concentrated on the wave properties. He tried to find a function that would encompass all informatin about a particle or system at any given time.

This is called the wave function. While debated ensued about which was correct, Schoedinger proved shortly after that the matrix mechanics and the wave function were mathematically equivalent.

Uncertainty Principle

Heisenberg then came up with the Uncertainty Principle, which stated that the position and velocity(momentum, which is mass*velocity) cannot be known precisely simultaneouely. The more accurately one is known, the less accurately the other can be measured.

The product of the uncertainty in position, and the uncertainty in momentum, is always greater than or equal to h/(2pi). While this is neglegible on the macroscopic level, subatomically, it causes great uncertainty. As a result, the world exists in statistical probabilities.

For example, a electron cannot be found precisely at one location. At any given chance, it has a certain probability of being found anywhere in the universe, with higher probabilities in regions around the atomic nucleus and less likelihood elsewhere. This can be calculated with the wave function.

Quantum Atomic Model

The quantum model of the atom calls for arbitrarily defined orbitals. An orbital is a region around the nucleus where an electron has 90% chance of being found.

Electrons have spin, a term used for the intrinsic angular momenta subatomic particles have. They have either a spin of +1/2 or -1/2. Two electrons in the same orbital of the same energy level must have opposite spin (Pauli Exclsion Principle).

Schroedinger's Cat

Since the wave function describes the world in terms of statistical probabilities and not certainties, a cat placed in an isolated system with equal chances of being found alive and dead at anytime *is* equally dead and alive at the same time. When an observer is introduced to the isolated system, the wave function collapses and the cat is either dead or alive.

PRINCIPLES OF QUANTUM MECHANICS

UNCERTAINTY PRINCIPLE

One of the most important concepts of quantum mechanics is the uncertainty principle, which states that there is always some uncertainty when trying to measure the posistion and momentum of a particle. More specifically, the product of the uncertainty in position, D *x*, and the uncertainty in momentum, D*mu*, is always greater than or equal to *h* over 2p, where *h* is planck's constant.

When trying to determine the position of an electron, for example, we cannot know *exactly* where it is and where it's going at the same time. The more accurately one is measured, the less accurately the other is known.

To we measure the position of the electron, for example, we can only obtain the information through the collision of the electron with another subatomic particle, such as a photon. After the collision, the momentum of the photon changes—it "bounces" away, and from this change in momentum the position of the electron is calculated.

However, the photon also gives the electron a change in momentum, so that now the position and velocity of the electron has changed from what it was. Thus, we can't know precisely the position and velocity simultaneously.

ELECTRON IN THE BOHR ATOM

In the Bohr model, electrons revolved around the nucleus in orbits. According to classical physics, they should radiate energy as they move and ultimately fall to the centre of the atom. Yet this is not so. The Quantum explanation is that electrons can only exist in predefined, discrete energy levels, and nowhere in between.

The lowest energy level is not at the centre of the atom, and electrons may go no closer than the lowest level. This explains why they don't collide into the nucleus. To move to another energy level, they must gain or lose *exactly* the energy difference between the levels. This explains spectral lines. Another explanation is that, if electrons crashed into the nucleus, their velovity would be zero and the uncertainty would also be zero. This would be a direct violation of the uncertainty principle, a fundamental property of nature.

SPIN AND QUANTUM NUMBERS

It was discovered experimentally that subatomic particles have intrinsic angular momentum. That is, that angular momentum is not given by some interaction, but a property of the particles.

This is a quantum property that classical physics has no explanation for. The closest thing on the macroscopic level is rotation (since angular momentum leads to rotating), so this property is called "spin". Subatomic particles, however, don't really spin; they just have the angular momenta. This property is quantatized since the particles may only have certain discrete values of spin and not any value in between. Electrons have spins of either +1/2 or −1/2.

In the quantum model, electrons in the atom are given four quantum numbers to distinguish them. The first three distinguish the energy level, the shape of the orbital, and in which orbital the electron is located (has 90% chance of being found), and the last one gives the spin.

The Pauli Exclusion Principle states that no 2 electrons in the atom can have the same 4 quantum numbers. If they occupy the same energy level and orbital, then they must have opposite spin.

NUCLEUS

The nucleus is a compact and dense region in the centre of the atom that makes up for most of its mass. It is made up of a group of subatomic particles known collectively as nucleons. The two states of nucleons are protons and neutrons.

Although the electron, and not the nucleus, is responsable for chemical reactions, the nucleus still holds much information about the atom. Since nucleons are about 2000 times heavier than electrons, the mass of the nucleus is approximately equal to the mass of the atom.

The number of protons in the nucleus determines the type of atom, and nuclear charge is the property by which the elements are arranged in the periodic table.

NUCLEAR ENERGY

A tremendous amount of energy is used to hold the nucleus together, since the repulsion between the positively charged protons is great. The bonding forces that hold the nucleus together are collectively known as the nucleur forces. When heavy element nuclei are split into smaller nuclei, or light element nuclei combined to make one nucleus, tremendous energies are released. The former is known fission, and the latter, fusion, and they are the driving forces behind stars and nuclear weopons.

HISTORY

From Dalton's theory, it was thought that atoms were indivisible particles that made up everything in the world. When Rutherford made his gold foil experiment, however, he found that when high-velocity positive alpha particles bombarded a thing metal foil, while most passed straight through, some were reflected back or deflected.

This led him to postulate that most of the atom was empty space, with most of the mass concentrated in one dense region in the centre of the arom. Thus, most of the positive alpha particles passed through the empty space, but some hit the dense core and were reflected back by electric repulsion. He called this the nucleus of the atom.

Subsequently, improvements were made on the Rutherford model. The nucleu was theorized to contain both positive and neutral particles.

APPLICATION

The interactions in the nucleus, the strong and the weak forces, are the strongest forces in nature. Processes such as fission and fusion release much of this energy, and make possible new energy sources. In fact, fusion is the process by which stars generate light and heat, and thus the source from which all life on earth is derived. Other processes, such as radiation, have medical applications.

STRUCTURE OF NUCLEUS

The Dalton atomic model suspected no nucleus. It assumed that atoms were the fundamental building blocks of nature and no particles exist that are smaller than atoms. However, Rutherford's Gold-foil experiment led him to suspect that atoms had nuclei where most of its mass was located, and that the nucleus was positive.

Bombarding Nitrogen with alpha particles, he noticed that the disintegration of the nucleus produced a positive particle equal to the mass of a hydrogen atom. He called this particle the proton and postulated that they made up the nucleus.

But a lot of positive charge concentrated in a small region would cause a lot of repulsion forces. So when Chadwick discovered the neutron in 1932, Heisenberg created the proton-neutron model of the nucleus. It also successfully explains isotopes.

THE PROTON

The proton is the nucleon that has a positive unit chage. Its mass is one atomic mass unit (amu, or 1.67×10^{-27} kg, 1,836 times the mass of an electron). It is present in the nucleus of every element.

Protons are responsible for the nuclear charge, the positive charge of the nucleus that deflected Rutherford's alpha particles. This charge also corresponds to the atomic number, which determines what kind of element the atom belongs to, and where on the periodic table the element belongs.

THE NEUTRON

The neutron is an electrically neutral particle slightly heavier than the proton. They are present in numbers greater than or equal to the number of protons in every element except Hydrogen (whose nucleus is a single proton), and variations in the number of neutrons of elements accound for isotopes.

The neutron has a maganetic moment, which suggests that its internal structure is made up of electrically charged components whose net charge is zero. The neutron is actually the fusion of a proton and an electron (since in beta decay a neutron is turned into a proton and an electron) and a neutrino.

ISOTOPES

Isotopes are atoms of the same element that have differenet mass. This is explained by the proton-neutron model of the nucleus. The number of protons account for which element it is and thus how it behaves, but the number of neutrons there are determines which isotope it is, and there can be more neutrons than there are protons.

For example, elemental Hydrogen has one proton (H-1) and a mass of one amu. But an isotope of Hydrogen, deuterium, has a proton and a neutron, and a mass of two amu. As another example, Carbon-12, Carbon-13 and Carbon-14 all have 6 protons (this determines that it is Carbon atom) but have 6, 7 and 8 neutrons, respectively.

Sometimes there are too many neutrons in the nucleus. When this happens, the neutron goes under beta decay, a type of radioactive decay.

The neutron spiits into a proton, an electron (beta particle) and a neutrino. The proton stays in the nucleus, but the beta particle and the neutrino(an electrically neutral, light or massless particle) are emitted, and energy is released, carried by the electron (when the electron doesn't carry all the energy, the rest is carried by the neutrino).

CLASSIFICATION

Nucleons are generally knoen as a baryon. Baryons are elementary particles consisting of three quarks and no antiquarks. They obey the Fermi-Dirac statistics, which encompass all elementary particles that obey the Pauli Exclusion Principle. The proton is the lightest baryon and is made of two up-quarks and one down-quark. The neutron is made of two down-quarks and one up-quark.

Nucleon's antiparticles are the antiproton and the antineutron, which have the same mass as the corresponding nucleons but opposite electric charge and magnetic moment.

TYPES OF RADIOACTIVITY

The study of radioactive material started with Becqueral in 1896, when he discovered that some of his films became unuseable, as if they had been exposed to light, even though they were kept in a sealed opaque container. Further investigation revealed a substance that gave a green glow in the dark.

This was uranium, and Becqueral discovered the X-ray. Marie Curie then studied radioactive material, discovering several radioactive substances. This research led to a better understanding of radiation, and peopls started bombarding elements with particles emitted from radiation. Rutherford's experimentations, for example, led him to his atomic model. These bombardments also led to the discovery of many particles, such as the proton. The nature of radiation was also better understood as the identity of the radiated particles came to be known, and theories of radioactive decay were devised.

These experiments did not reveal a complete picture of radiation, until Fermi's systematic bombardment of all elements. The results led him to envision a chain reaction, where the bombardment of an atom led to its splitting and emitting more particles that bombarded the other atoms. This was carried out in 1942, and led to the development of the A-bomb shortly after.

ALPHA PARTICLES

Alpha radiation (or alpha rays) was distinguished and named by E. R. Rutherford in 1909, who found by measuring the charge and mass of alpha particles that they are the nuclei of ordinary helium atoms. Alpha particles consist of two protons and two neutrons.

BETA PARTICLES

Beta radiation (or beta rays) was identified and named by E. Rutherford too. He found that beta rays consists of high-speed electrons. Unlike alpha and gamma particles, whose energy can be explained as the difference of the energies of the radioactive nucleus before and after emission, beta particles emerge with a variable energy.

This apparent violation of the law of conservation of energy led to the hypothesis that a second undetected particle, the neutrino, is emitted along with the electron and shares the total available energy. In some forms of induced, or artificial, radioactivity, the electron's antiparticle, the positron, is emitted from the excited nucleus; the positron in this case is also called a beta particle and denoted by b^+ (the ordinary beta particle is b^-).

GAMMA RADIATION

Gamma radiation is high-energy photons emitted as one of the three types of radiation resulting from natural radioactivity. It is the most energetic form of electromagnetic radiation, with a very short wavelength (high frequency). Wavelengths of the longest gamma radiation are less than 10^{-10} m, with frequencies greater than 10^{18} hertz (cycles per sec). Gamma rays are essentially very energetic X rays; the distinction between the two is not based on their intrinsic nature but rather on their origins.

X rays are emitted during atomic processes involving energetic electrons. Gamma radiation is emitted by excited nuclei or other processes involving subatomic particles; it often accompanies alpha or beta radiation, as a nucleus emitting those particles may be left in an excited (higher-energy) state. The applications of gamma radiation are much the same as those of X rays, both in medicine and in industry. In medicine, gamma ray sources are used for cancer treatment and for diagnostic purposes.

Some gamma-emitting radioisotopes are also used as tracers. In industry, principal applications include inspection of castings and welds.

Data from artificial satellites and high-altitude balloons have indicated that a flux of gamma radiation is reaching the earth from outer space, thus opening up the field of research known as gamma-ray astronomy.

DECAY

ALPHA DECAY

There are three types of radioactive decay, named by the type of particle they produced. All radioactive decay is the result of an unstable nucleus and release energy. The first is the alpha decay, which emits an alpha particle, a positively charged and heavy particle that does not penetrate too well.

It was shown that the alpha particle is actually a Helium nucleus, consisting of two protons and two neutrons. When the unstable nucleus emits the alpha particle, the atomic number of the atom decreases by two and the atomic mass number decreases by four.

BETA DECAY

The beta particle is an electron. When beta decay occurs, a neutron in the nucleus converts into a proton and an electron. The electron is emitted and takes the released energy with it. When it does not carry enough enery, however, the remaining is carried by a small, mass less (or very light) neutral particle called the neutrino. The beta particle is therefore negative and more penetrating than the alpha particle.

GAMMA DECAY

The Gamma radiation is unaffected by magnetic field as it travels; therefore it is electrically neutral. It is actually a type of electromagnetic radiation, and not a particle of significant mass.

Gamma rays is one of the most energetic types of electromagnetic radiation, because its wavelength is very short and its frequency very high, (they are therefore invisible to the eye). Gamma radiation often accompanies alpha decay or beta decay, since the nucleus emitting those particles may still be excited, and the gamma ray is the result of the extra energy. Gamma rays are the most penetrating type of radiation.

NATURAL GROWTH

Radioactive decay obeys the rules of natural (exponential) growth or decay. That is, the rate of decay is directly proportional to the amount of undecayed substance present. (Chemically, the rate of reaction is directly proportional to the amount of unreacted reactant.) Thus, the first derivative of the substance, say, Uranuim (U) with respect to time is proportional to the amount of undecayed U present

d[U]/dt = k[U].

Solving this gives d[U]/[U] = kdt, or

ln[U] + C = kt, [U] = a*e(kt),

where a is calculated to be the amounf or Uranium at the start of the decay.

A proporty of the exponential graph is that starting at any arbitrarily chosen time, the time it takes for the amount of undecayed Uranium to decrease by half (or any fraction) is the same.

This period of time, the time it take for half the sample to undergo radioactive decay, is called the Half-life of that element. After one half-life, the amount of unreacted sample left is alway one half the amount of unreacted sample before that period of time.

Thus, if the half life of a particular sample is 10 years, then after 20 years one-quarter of undecayed sample will be left.

INDUCED RADIOACTIVITY

The induced radioactivity was discovered by French physicists Iren Joliot-Curie (1897-1956) and Frederic Joliot-Curie (1900-1960). The first synthetic nuclear reaction was realized by Rutherford. It is possible to register the charged particles which are taking off at a radioactive decay. It allows to use atoms of the synthetic radioactive isotopes.

NUCLEAR REACTIONS

Nuclear reactions. There are 2 stages of nuclear reactions. The first stage of nuclear reaction consists in entrapment by a kern of a particle. For a small time interval the energy is uniformly proportioned between all particles of a compound nucleus.

The second stage of nuclear response (start of a particle from a compound nucleus) occurs as a result of casual diversions from a uniform distribution of energy. There is a time between entrapment of particles and their start from a kern, which is significant more then nuclear time $\tau_n = 10^{-22}$ s. All process of nuclear reaction is possible to figure by the formula:

$${}_{Z1}X^{A1} + a \rightarrow {}_{Z2}Y^{A2} \rightarrow {}_{Z3}C^{A3} + b$$

${}_{Z1}X^{A1}$ -inital kern *a, b* particles

${}_{Z2}Y^{A2}$ -compound nucleus; ${}_{Z3}C^{A3}$ -terminating kem

DIVISION OF KERNS

The heavy kerns are labile. The kern, which was exited as a result of a neutron capture, can be divided into identical parts. The nuclear fission on two debrises is accompanied by eduction of huge energy.

The debrises, which are formed at division, are radioactive and can emit neutrons. As the result the exuberant number of neutrons is formed, that gives in radio-activity of yields of decay.

REACTION OF DIVISION

CHAIN REACTION OF DIVISION

The division of kerns of uranium occurs at interaction to prompt or sluggish neutrons, but the irradiation by sluggish neutrons is more effective. The practical interest is represented by reactions, at which the major energy is oozed and there are neutrons, which give in magnification of number of the acts of division. For realization of chain reaction the maintenance of fixed number of neutrons is necessary.

For an estimation of number of neutrons is inlet coefficient of manifolding of neutrons k. It determines the ratio of number of neutrons, which arise at reaction, to number of neutrons, which participated in reaction. If $k >= 1$, the reaction will be explicated.

The minimum sizes of a fissile region, at which the realization of a chain reaction is possible, term as the critical sizes. The minimum quantity of a fissionable material necessary to sustain a nuclear chain reaction is called the critical mass.

CONTROLLABLE REACTION OF DIVISION

The controllable reactions occurs in a spacial nuclear reactors. Reactors can be used for research or for power production. A research reactor is designed to produce various beams of radiation for experimental application; the heat produced is a waste product and is dissipated as efficiently as possible.

In a power reactor the heat produced is of primary importance for use in driving conventional heat engines; the beams of radiation are controlled by shielding.

A-BOMB

Practical fissionable nuclei for atomic bombs are the isotopes uranium-235 and plutonium-239, which are capable of undergoing chain reaction. If the mass of the fissionable material exceeds the critical mass (a few pounds), the chain reaction multiplies rapidly into an uncontrollable release of energy.

An atomic bomb is detonated by bringing together very rapidly (e.g., by means of a chemical explosive) two sub critical masses of fissionable material, the combined mass exceeding the critical mass.

An atomic bomb explosion produces, in addition to the shock wave accompanying any explosion, intense neutron and gamma radiation, both of which are very damaging to living tissue. The neighborhood of the explosion becomes contaminated with radioactive fission products. Some radioactive products are borne into the upper atmosphere as dust or gas and may subsequently be deposited partially decayed as radioactive fallout far from the site of the explosion.

ELEMENTARY PARTICLES

All subatomic particles, such as nucleons or electrons, and the particles they release when undergoing processes such as radioactive decay, are organized into classes.

All particles that obey the Pauli Exclusion Principle (no two particles can have the same four quantum numbers in a system, such as an atom) are called fermions because the Fermi-Dirac statistics apply to them. All particles that don't obey the principle are called bosons because the Bose-Einstein statistics apply to them.

FERMIONS

Fermions have half-integral spin (intrinsic angular momentum). They are deemed anitsocial particles because they obey the exculsion principle, meaning that no two fermion with the same spin can be found in a system. The antiparticles for fermions have the same mass but opposite charge and magnetic moment as the corresponding particle.

BARYONS

The first type of fermion is the baryon. They are those particles that experience the strong nuclear force. They consist of three quarks (quark triplet). The most notable baryons are the nucleons, the two states of which are protons and neutrons, the particles that make up the nucleus.

Other, heavier baryons, called hyperons also exist, but they are unstable, and usually decays to a form of nucleon(heavier baryons include the sigma, the lamda and the delta particles). Since all baryons decay to the nucleon form, the nucleons are known as the baryon ground state.

All baryons obey the law of baryon family numbers (the family number is +1 for particles such as protons and -1 for antiparticles such as antiprotons), which states that the sum of the family number of all reactants in an interaction is equal to the sum of the family number of all products.

This means that, for every particle created or destroyed, a corresponding antiparticle is also created or destroyed. For example, a proton colliding with an antiproton will result in the mutual destruction of both particles, or, a proton and an antiproton can be created simultaneously.

LEPTONS

Leptons are fermions that do not experience the strong nuclear force. They are only weakly interactive, and can be created by nuclear decay. Leptons include such particles as electrons, muons, neutrinos and the tau particle.

So, for example, an electron and a neutrino can be created by the decay of an unstable neutron to a proton. As fermions they obey the exclusion principle, so no two electrons can occupy the same energy level and orbital of an atom (a system) and have the same spin. Leptons are the lightest fermions.

There are three separate laws concerning the conservation of lepton family numbers (the family numbers for leptons, like for baryons, is +1 for particles and -1 for antiparticles), for the electrons, muons and taus.

BOSONS

Bosons are particles that, unlike the fermions, do not obey the exclusion principle. Generally, fermions are particles that compose atomic structure but bosons carry out the interaction (forces) between fermions. Bosons have integral spin and are deemed social because many bosons can by placed in the same system in the same energy state at once.

MESONS

Mesons are a class of bosons whose mass falls between the lighter lepton fermions and the heavier baryon fermions. They consist of two quarks (quark pair), and include such particles as the p meson (pion) and the *K* meson (kaon). It was previously thought that the muon was a type of meson (m meson) because of its large mass (about 200 times that of the electron), but the behaviour of the muon doesn't fit into the description of mesons, so it was reclassified.

The pion is accountable for carrying out the strong nuclear force, which is the force that holdes the nucleus together, despite tmememndous electric repulsion.

Chapter 5

Radioactivity

Radiation is all around us. It comes from the Earth and from outer space. Many forms of radiation are invisible — we can't feel it, see it, taste it, or smell it. Yet, it can be detected and measured when present. We measure ionizing radiation in units called millirems. But what is radiation? Radioactive materials are composed of atoms that are unstable. An unstable atom gives off its excess energy until it becomes stable. The energy emitted is radiation. We can classify radiation as being either natural or man-made.

The Earth is surrounded by radiation. Every day, for example, we are exposed to radon, a radioactive gas from uranium found in soil dispersed in the air; from radioactive potassium in our food and water; from uranium, radium, and thorium in the Earth's crust; and from cosmic rays and the sun.

These types of radiation are called natural or background radiation. We are exposed to an average of 300 millirems of natural radiation each year (a millirem is a unit of measure for exposure to radiation). This amounts to natural radiation accounting for nearly 85 per cent of our total annual exposure. Where does the remaining 15 per cent come from? Man-made sources.

Man-made radiation sources that people can be exposed to include tobacco, television, medical x-rays, smoke detectors, lantern mantles, nuclear medicine, and building materials.Generally, when we think of exposure to radiation, we need to look at radioactive atoms produced in nuclear reactors and described as being unstable.

They are unstable because they undergo a disintegrating process called decaying. During this process, unstable atoms becomes stable, throwing off (emitting) radiation in the form of rays and/or particles.How fast a radioactive atom decays into a stable atom depends on the atom itself. For example, the range in the rate of decay among isotopes goes from fractions of seconds to several billion years (e.g., uranium).Let's take a look at uranium-238 to the decay chain.

$U_{238} \rightarrow$ $Thorium_{230} \rightarrow$ $Radium_{226} \rightarrow$ $Radon_{218} \rightarrow$ $Bismuth_{214} \rightarrow$ $Lead_{206}$

As U-238 decays it changes into thorium-230, which changes into radium-226, which changes into radon-218, which changes into bismuth-214, and finally into lead-206 (a stable element).

One peculiar thing about radioactive atoms is that no one knows exactly when the element will decay and give off radiation. There is, however, a pattern relating to how long it takes for an isotope to lose half of its radioactivity.

The pattern is called half-life. If an atom, for example, has a half-life of 10 years, half of its atoms will decay in 10 years. Then in another 10 years half of that amount will decay and so on. While there are several different forms of radiation, we're going to concentrate on just three that result from the decay of radioactive isotopes: alpha, beta, and gamma.

Beta particles are high energy electrons. Both alpha and beta particles are emitted from unstable isotopes. The alpha particle, consisting of two protons and two neutrons, is relatively large compared to beta particles. Gamma rays have no mass.

Because of its size and electrical charge (+2), the alpha particle has a relatively slow speed and low penetrating distance (one to two inches in air). Alpha particles are easily stopped by a thin sheet of paper or the body's outer layer of skin. Since they do not penetrate the outer (dead) layer of skin, they present little or no hazard when they are external to the body.

However, alpha particles are considered internal hazards, because when they come into contact with live tissue they cause a large number of ionizations to occur in small areas, thus causing damage to tissues and cells.

Beta radiation, while faster and lighter than alpha radiation, can travel through about 10 feet of air and penetrate very thin layers of materials such as aluminum foil. However, while clothing will stop most beta particles, they can penetrate the live layers of skin tissue.

Therefore, beta radiation is considered to be both an internal and external (to skin only) hazard. Thin layers of metals and plastics can be used to shield individuals from beta radiation.

Gamma radiation, high energy light, is a little different. It is a type of electromagnetic wave, just like radio waves, light waves, and x-rays.

Gamma radiation is a very strong type of electromagnetic wave. It is has no weight and travels at the speed of light. This is much faster than alpha and beta radiation.

Because of their penetrating capability, gamma rays are considered both internal and external hazards. Thick walls of cement, lead, or steel are needed to stop it.

Alpha, beta, and gamma radiations are also known as ionizing radiation. Ionizing radiation is especially harmful because it can change the chemical makeup of many things, including the delicate chemistry of the human body and other living organisms. For this reason it is a good idea to avoid unnecessary exposure to all ionizing radiation.

The problem is simply this: large amounts of radiation — far above the levels encountered in daily life — can produce cancer and genetic defects in living organisms. Radiation causes damage and alters the body's normal cells and normal cell function. This breakdown in normal cell function may result in an uncontrolled growth of cells, hence the potential for malignant/cancerous tumors.

Whether the source of radiation is natural or man-made, whether it is a small dose of radiation or a large dose, there will be some biological effects. A diagram can show the biological effect of ionizing radiation. Radiation causes ionizations of atoms that will affect molecules that may affect cells that may effect tissues that may affect organs that may affect the whole body.

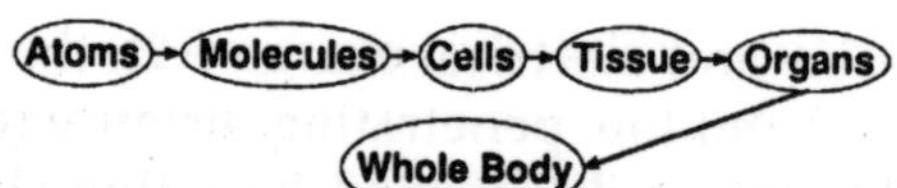

Although we tend to think of biological effects in terms of the effect of radiation on living cells, in actuality, ionizing radiation, by definition, interacts only with atoms by a process called ionization. Thus, all biological damage effects begin when radiation interacts with atoms forming the cells in the human body. As a result, radiation effects on humans proceed from the lowest to the highest level as noted on the board.

Although scientists have only known about radiation since the 1890s, they have developed a wide variety of uses for this natural force. Today, to benefit mankind, radiation is used in science, medicine, and industry, as well as for generating electricity. Radiation has useful applications in such areas as agriculture, medicine, space exploration, architect/ engineering, industry/manufacturing, government, geology (including mining), ecology, and education.

Radiation is used by doctors to diagnose illness and helps archaeologists find the age of ancient artifacts. Electricity produced by nuclear fission — splitting the atom — is one of its greatest uses. A reliable source of electricity is needed to give us light, help to groom and feed us, and to keep our homes and businesses running. Let me give you some specific examples of how the radiation has been used to:

- Diagnose and treat illnesses.
- Kill bacteria and preserve food without chemicals and refrigeration.
- Process sludge for fertilizer and soil conditioner.
- Locate underground natural resources and tell a dry hole from a gusher.
- Make smoke detectors, nonstick fry pans, and ice cream.
- Grow stronger crops.
- Power satellites and provide future electrical needs for space laboratories with people on board.
- Design instruments, techniques, and equipment; measure air pollution.
- Prove the age of works of art and assist in determining their authenticity.

RADIATION IN SCIENCE

Radiation is used in science in many ways. Just as doctors can label substances inside people's bodies, scientists can label substances that pass through plants, animals, or our world. This allows us to study such things as the paths that different types of air and water pollution take through the environment.

It has also helped us learn more about a wide variety of things, such as what types of soil different plants need to grow, the size of newly discovered oil fields, and the track of ocean currents.

Scientists use radioactive substances to find the age of ancient objects by a process called carbon dating. For example, in the upper levels of our atmosphere, cosmic rays hit nitrogen atoms and form a naturally radioactive isotope called carbon-14. Carbon is found in all living things, and a small percentage of this carbon is carbon-14. When a plant or animal dies, it no longer takes in new carbon and the carbon-14 it contains begins the process of radioactive decay.

However, new isotopes of carbon-14 continue to be formed in our atmosphere, and after a few years the per cent of radioactivity in an old object is less than it is in a newer one. By measuring this difference, scientists are able to determine how old certain objects are. The measuring process is called carbon dating.

RADIATION USED TO SOLVE CRIMES

How is radiation used to solve crimes? I am sure that you already know that detectives often search the scene of a crime for traces of paint, glass, hair, gunpowder, or blood. But you may not know that after such evidence is collected, it is often exposed to radiation and then analyzed to find out its exact makeup.

If material is exposed to streams of neutrons, some of the neutrons can be absorbed into the nucleus of the exposed material. This makes these materials slightly radioactive because they are unstable and decay with time. Scientists are then able to read the exact chemical signatures of these substances. This laboratory process, called activation analysis, is precise enough to tell if a single hair found at a crime scene came from a certain person.

Activation analysis is also used to find out the chemical makeup of materials when scientists only have small samples, as well as to prove that older works of art are not made of modern materials.

RADIATION IN INDUSTRY

We could talk all day about the many and varied uses of radiation in industry and not complete the list. To make a long story short, we'll just concentrate on a few. Exposure to some types of radiation (for example, x-rays) can kill germs without harming the items that are being disinfected or making them radioactive.

For example, when treated with radiation, foods take much longer to spoil, and medical equipment such as bandages, hypodermic syringes, and surgical instruments don't have to be exposed to toxic chemicals or extreme heat to be sterilized. Although we now use chlorine, a toxic and difficult-to-handle chemical, we may use radiation in the future to disinfect our drinking water and even kill all the germs in our sewage. Ultraviolet light already is used to disinfect drinking water in some homes.

The agricultural industry makes use of radiation to improve food production. Plant seeds, for example, have been exposed to radiation to bring about new and better types of plants. Besides making plants stronger, radiation can also be used to control insect populations, thereby decreasing the use of pesticides.

Engineers use radioactive substances to measure the thickness of materials and an x-ray process called radiography to find hard to detect defects in many types of metals and machines. Radiography is also used to check such things as the flow of oil in sealed engines and the rate and way various materials wear out.

And we've already talked about the use of the radioactive element uranium, which is used as a fuel to make electricity for our cities, farms, towns, factories, etc.

In outer space, radioactive materials are also used to power space craft. Such materials have also been used to supply electricity to satellites sent on missions to the outermost regions of our solar system.

Radiation has been used to help clean up toxic pollutants, such as exhaust gases from coal-fired power stations and industry. Sulphur dioxides and nitrogen oxides, for example, can be removed by electron beam radiation.

As you can see, radiation and radioactive materials have played and will continue to play a very significant role in our lives. Let's sum up this discussion with a walk through the life of a typical family for one day and learn about some of the uses of radiation.

Dad gets up in the morning and puts on a clean shirt. His polyester-cotton blend shirt is made from chemically treated fabric that has been irradiated (treated with radiation) before being exposed to a soil-releasing agent. The radiation makes the chemicals bind to the fabric, keeping his shirt fresh and pressed all day. The shirt is not radioactive.

In the kitchen, Jenny is frying an egg. That nonstick pan she is using has been treated with gamma rays, and the thickness of the eggshell was measured by a gauge containing radioactive material before going into the egg carton. Thin, breakable eggs were screened out. The turkey Mom is taking out of the refrigerator for tonight's dinner was covered with irradiated polyethylene shrink wrap. Once polyethylene has been irradiated, it can be heated above its usual melting point and wrapped around the turkey to provide an airtight cover.

As Dad drives to work, he passes reflective signs that have been treated with radioactive tritium and phosphorescent paint. During lunch, brother Bob has some ice cream. The amount of air whipped into that ice cream was measured by a radioisotopic gauge.

After you and your family return home this evening, some of you may have soda and others may sit and relax. Nuclear science is at work here: The soda bottle was carefully filled — a radiation detector prevented spillover. And your family is safe at home because the ionizing smoke detector, using a tiny bit of americium-241, will keep watch over you while you sleep.

Increasingly, our country has become a nation of electricity users. We depend on an abundant affordable supply of energy to power the many machines we use in our complex society. Can you imagine what it would be like not having electricity in your home? About one-third of our energy resources are used to produce electricity.

Electricity can be produced in many ways — most of which you already know about. Today, we're going to talk about one of those ways

— nuclear fission. In America, nuclear energy plants are the second largest source of electricity after coal — producing approximately 21 per cent of our electricity. There is something I want all of you to be aware of: The purpose of a nuclear power plant is to produce electricity.

While nuclear power plants have many similarities to other types of electricity generating plants, there are some significant differences.

With the exception of solar, wind, and hydroelectric plants, all others including nuclear convert water to steam that spins the propeller-like blades of a turbine that spins the shaft of a generator. Inside the generator coils of wire and magnetic fields interact to create electricity.

The energy needed to boil water into steam is produced in one of two ways: by burning coal, oil, or gas (fossil fuels) in a furnace or by splitting certain atoms of uranium in a nuclear energy plant. Nothing is burned or exploded in a nuclear energy plant. Rather, the uranium fuel generates heat through a process called fission.

Uranium is an element that can be found in the crust of the Earth. This element, quite abundant in many areas of the world, is naturally radioactive. Certain isotopes of uranium can be used as fuel in a nuclear power plant.

The uranium is formed into ceramic pellets about the size of the end of your finger. These pellets are inserted into long, vertical tubes (fuel rods) within the reactor. The reactor is the heart of the nuclear power plant. Basically, it is a machine that heats water.

A reactor has four main parts: the uranium fuel assemblies, the control rods, the coolant/moderator, and the pressure vessel. The fuel assemblies, control rods, and coolant/moderator make up what is known as the reactor core. The core is surrounded by the pressure vessel.

We also have to understand that uranium cannot just be thrown into a reactor the way we shovel coal into a furnace. The fuel rods, containing the uranium, are carefully bound together into fuel assemblies, each of which contains about 240 rods. The assemblies hold the rods apart so that when they are submerged into the reactor core, water can flow between them.

When the uranium atom splits, it releases energy and two or more neutrons from its nucleus. These neutrons can then hit the nuclei of other uranium atoms causing them to fission. The sequence of one fission triggering others, and those triggering still more, is called a chain reaction. When the atoms split, they release energy in the form of heat.

The heat is transferred from the reactor core to the steam generator. Here, it's converted to high pressure steam which turns the turbine in the electric generator.

This next transparency shows the basic design of the nuclear fuel steam plant and its various components. In the past few minutes I've introduced several new terms. Let me explain what each means in the energy generating process. First, control rods. What are they? How are they used?

The control rods slide up and down in between the fuel assemblies in the reactor core. They control or regulate the speed of the nuclear reaction by absorbing neutrons. Here's how it works: When the control rods absorb neutrons, fewer neutrons hit the uranium atoms thus slowing down the chain reaction. On the other hand, when the core temperature goes down, the control rods are slowly lifted out of the core, and fewer neutrons are absorbed. Therefore, more neutrons are available to cause fission. This releases more heat energy.

Just as there are different types of houses and cars, there are different types of nuclear power plants that generate electricity. The two basic types being used are the boiling water reactor (BWR) and the pressurized water reactor (PWR). These power plants are often referred to as light water reactors.

BOILING WATER REACTOR (BWR)

The boiling water reactor operates in essentially the same way as a fossil fuel generating plant. Neither of these types of power plants have a steam generator. Instead, water in the BWR boils inside the pressure vessel and the steam water mixture is produced when very pure water (reactor coolant) moves upward through the core absorbing heat. The water boils and produces steam. When the steam rises to the top of the pressure vessel, water droplets are removed, the steam is sent to the turbine generator to turn the turbine.

PRESSURIZED WATER REACTOR (PWR)

The pressurized water reactor differs from the BWR in that the steam to run the turbine is produced in a steam generator. Water boils at 212°F or 100°C. If a lid is tightly placed over a pot of boiling water (a pressure cooker), the pressure inside the pot will increase because the steam cannot escape. As the pressure increases, so does the temperature of the water in the pot.

In the PWR plant, a pressurizer unit keeps the water that is flowing through the reactor vessel under very high pressure to prevent it from boiling. The hot water then flows into the steam generator where it is converted to steam. The steam passes through the turbine which produces electricity.

Chapter 6

Atomic Structure

DISCOVERY OF α, β AND y PARTICLES

The discovery of x-rays by William Conrad Roentgen in November of 1895 excited the imagination of a generation of scientists who rushed to study this phenomenon. Within a few months, Henri Becquerel found that both uranium metal and salts of this element gave off a different form of radiation, which could also pass through solids.

By 1898, Marie Curie found that compounds of thorium were also "radioactive." After pain-staking effort she eventually isolated two more radioactive elements polonium and radiumfrom ores that contained uranium.

In 1899 Ernest Rutherford found that there were at least two different forms of radioactivity when he studied the absorption of radioactivity by thin sheets of metal foil.

One, which he called alpha (α) particles, were absorbed by metal foil that was a few hundredths of a centimeter thick. The other, beta (β) particles, could pass through 100 times as much metal foil before they became absorbed. Shortly thereafter, a third form of radiation, gamma (γ) rays, was discovered that could penetrate as much as several centimeters of lead.

The direction in which -particles were deflected by an electric field suggested that they were positively charged. The magnitude of this deflection suggested that they had the same charge-to-mass ratio as an He^{2+} ion.

To test the equivalence between -particles and He^{2+} ions, Rutherford built an apparatus that allowed -particles to pass through a very thin glass wall into an evacuated flask that contained a pair of metal electrodes. After a few days, he connected these electrodes to a battery and noted that the gas in the flask did indeed give off the characteristic emission spectrum of helium.

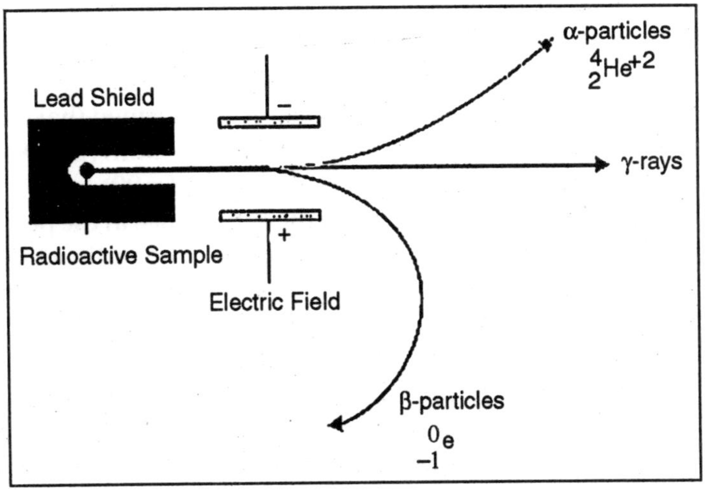

Fig. The Effect of an Electric F ield on -, b, and -Radiation.

Experiments with electric and magnetic fields demonstrated that β particles were negatively charged. Furthermore, they had the same charge-to-mass ratio as an electron. To date, no detectable difference has been found between β-particles and electrons. The only reason to retain the name " β-particle" is to emphasize the fact that these particles are ejected from the nucleus of an atom when it undergoes radioactive decay.

The fact that -rays are not deflected by either electric or magnetic fields suggests that these rays don't carry an electric charge. Since they travel at the speed of light, they are classified as a form of electromagnetic radiation that carries even more energy than x-rays.

GROWTH AND DECAY CURVES

At the turn of the century, when radioactivity was discovered, atoms were assumed to be indestructible. Ernest Rutherford and Frederick Soddy, however, found that radioactive substances became less active with time. More importantly, they noticed that radioactivity was always accompanied by the formation of atoms of a different element. By 1903, they concluded that radioactivity was accompanied by a change in the structure of the atom.

They therefore assumed that radiation was emitted when an element decayed into a different kind of atom.By 1910, 40 radioactive elements had been isolated that were associated with the process by which uranium metal decayed to lead. This created a problem, however, because there was space for only 11 elements between lead and uranium. In 1913, Kasimir Fajans and Fredick Soddy proposed an explanation for these results based on the following rules.

- α -particles are emitted when an element is formed that belongs two spaces to the left in the periodic table. Uranium (Z = 92), for example, emits an -particle when it decays to form thorium (Z = 90).
- β -particles are emitted when an element is formed that belongs one space to the right in the periodic table. Actinium (Z = 89) emits a β -particle when it decays to form thorium (Z = 90).
- Radioactive elements that fall in the same place in the periodic table are different forms of the same element. The radioactive thorium produced by the -particle decay of uranium is a different form of the element than the radioactive thorium obtained by the β -particle decay of actinium.

Soddy proposed the name isotope to describe different radioactive atoms that occupy the same position in the periodic table. J. J. Thomson and Francis Aston then used a mass spectrometer to show that isotopes are atoms of the same element that have different atomic masses.

THE STRUCTURE OF THE ATOM

The discovery of the electron in 1897 by J. J. Thomson suggested that there was an internal structure to the "indivisible" building blocks of matter known as atoms. This raised an obvious question: How many electrons does an atom contain? By studying the scattering of light, x-rays, and -particles, Thomson concluded that the number of electrons in an atom was between 0.2 and 2 times the weight of the atom.

In 1911, Rutherford concluded that the scattering of -particles by extremely thin pieces of metal foil could be explained by assuming that all of the positive charge and most of the mass of the atom were concentrated in an infinitesimally small fraction of the total volume of the atom, for which he proposed the name nucleus. Rutherford's data also suggested that the nucleus of a gold atom carries a positive charge that is about 80 times the charge on an electron.

The discovery of the neutron in 1932 explained the discrepancy between the charge on the nucleus and the mass of an atom. A neutral gold atom that has a mass of 197 amu consists of a nucleus that contains 79 protons and 118 neutrons surrounded by 79 electrons. By convention, this information is specified by the following symbol, which describes the only naturally occurring isotope of gold.

Mass Number → 197

Au

charge on NUCLEUS → 79

This convention can also be applied to subatomic particles. The only difference is the use of lowercase letters to identify the particle.

$$\begin{matrix} 0 & 1 & 1 \\ -1^{e} & 1^{p} & 0^{n} \end{matrix}$$

electron proton neutron

Because anyone with access to a periodic table can find the atomic number of an element, a shorthand notation is often used that reports only the mass number of the atom and the symbol of the element. The shorthand notation for the naturally occurring isotope of gold is ^{197}Au.

RADIOACTIVE DECAY

α, β and γ Radiation

Early studies of radioactivity indicated that three different kinds of radiation were emitted, symbolized by the first three letters of the Greek alphabet, β and. With time, it became apparent that this classification scheme was much too simple.

The emission of a negatively charged β particle, for example, is only one example of a family of radioactive transformations known as β -decay. A fourth category, known as spontaneous fission, also had to be added to describe the process by which certain radioactive nuclides decompose into fragments of different weight.

ALPHA DECAY

Alpha decay is usually restricted to the heavier elements in the periodic table. (Only a handful of nuclides with atomic numbers less than 83 emit an -particle). The product of -decay is easy to predict if we assume that both mass and charge are conserved in nuclear reactions. Alpha decay of the ^{238}U "parent" nuclide, for example, produces ^{234}Th as the "daughter" nuclide.

$$^{238}_{92}U \rightarrow {}^{234}_{90}Th + {}^{4}_{2}He$$

The sum of the mass numbers of the products (234 + 4) is equal to the mass number of the parent nuclide (238), and the sum of the charges on the products (90 + 2) is equal to the charge on the parent nuclide.

BETA DECAY

There are three different modes of beta decay:

- Electron (β^-) emission
- Electron capture
- Positron (β^+) emission

Electron (β^-) emission is literally the process in which an electron is ejected or emitted from the nucleus. When this happens, the charge on the nucleus increases by one. Electron (β^-) emitters are found throughout the periodic table, from the lightest elements (^3H) to the heaviest (^{255}Es). The product of β^--emission can be predicted by assuming that both mass number and charge are conserved in nuclear reactions. If ^{40}K is a β^--emitter, for example, the product of this reaction must be ^{40}Ca.

$$^{40}_{19}K \rightarrow \,^{40}_{20}Ca + \,^{0}_{-1}e$$

Once again the sum of the mass numbers of the products is equal to the mass number of the parent nuclide and the sum of the charge on the products is equal to the charge on the parent nuclide.

Nuclei can also decay by capturing one of the electrons that surround the nucleus. Electron capture leads to a decrease of one in the charge on the nucleus. The energy given off in this reaction is carried by an x-ray photon, which is represented by the symbol hv, where h is Planck's constant and v is the frequency of the x-ray. The product of this reaction can be predicted, once again, by assuming that mass and charge are conserved.

$$^{40}_{19}K + \,^{0}_{-1} \rightarrow \,^{40}_{18}Ar + hv \qquad \text{(Electron capture)}$$

The electron captured by the nucleus in this reaction is usually a 1s electron because electrons in this orbital are the closest to the nucleus.

A third form of beta decay is called positron (β^+) emission. The positron is the antimatter equivalent of an electron. It has the same mass as an electron, but the opposite charge. Positron (β^+) decay produces a daughter nuclide with one less positive charge on the nucleus than the parent.

$$^{40}_{19}K + \,^{40}_{18}Ar + \,^{0}_{+1}e \qquad \text{(Position } (\beta^+) \text{ emission)}$$

Positrons have a very short life-time. They rapidly lose their kinetic energy as they pass through matter. As soon as they come to rest, they combine with an electron to form two -ray photons in a matter-antimatter annihilation reaction.

$$^{0}_{+1}e + \,^{0}_{-1}e \rightarrow 2\gamma$$

Thus, although it is theoretically possible to observe a fourth mode of beta decay corresponding to the capture of a positron, this reaction does not occur in nature.

Note that in all three forms of β -decay for the ^{40}K nuclide the mass number of the parent and daughter nuclides are the same for electron emission, electron capture, and position emission. All three forms of β - decay therefore interconvert isobars.

GAMMA EMISSION

The daughter nuclides produced by -decay or β -decay are often obtained in an excited state. The excess energy associated with this excited state is released when the nucleus emits a photon in the -ray portion of the electromagnetic spectrum.

Most of the time, the -ray is emitted within 10^{-12} seconds after the - particle or β -particle. In some cases, gamma decay is delayed, and a short-lived, or metastable, nuclide is formed, which is identified by a small letter m written after the mass number. ^{60m}Co, for example, is produced by the electron emission of ^{60}Fe.

$$^{60}_{26}Fe \rightarrow ^{60}_{27}Co + ^{0}_{-1}e$$

The metastable ^{60m}Co nuclide has a half-life of 10.5 minutes. Since electromagnetic radiation carries neither charge nor mass, the product of -ray emission by ^{60m}Co is ^{60}Co.

$$^{60}_{26}Co \rightarrow ^{60}_{27}Co + \gamma \qquad (\gamma - \text{ray emission})$$

SPONTANEOUS FISSION

Nuclides with atomic numbers of 90 or more undergo a form of radioactive decay known as spontaneous fission in which the parent nucleus splits into a pair of smaller nuclei. The reaction is usually accompanied by the ejection of one or more neutrons.

$$^{252}_{98}Cf \rightarrow ^{140}_{54}Xe + ^{108}_{44}Ru + 4\,^{1}_{0}n$$

For all but the very heaviest isotopes, spontaneous fission is a very slow reaction. Spontaneous fission of ^{238}U, for example, is almost two million times slower than the rate at which this nuclide undergoes -decay.

NEUTRON-RICH VERSUS NEUTRON-POOR NUCLIDES

In 1934 Enrico Fermi proposed a theory that explained the three forms of beta decay. He argued that a neutron could decay to form a proton by emitting an electron. A proton, on the other hand, could be transformed into a neutron by two pathways. It can capture an electron or it can emit a positron. Electron emission therefore leads to an increase in the atomic number of the nucleus.

$$^{14}_{6}C \rightarrow ^{14}_{7}N + ^{0}_{-1}e$$

Both electron capture and positron emission, on the other hand, result in a decrease in the atomic number of the nucleus.

$$ {}^{7}_{4}Be + {}^{0}_{-1}e \rightarrow {}^{7}_{3}Li + hv $$

$$ {}^{11}_{6}C \rightarrow {}^{11}_{5}B + {}^{0}_{+1}e $$

A plot of the number of neutrons versus the number of protons for all of the stable naturally occurring isotopes. Several conclusions can be drawn from this plot.

- The stable nuclides lie in a very narrow band of neutron-to-proton ratios.
- The ratio of neutrons to protons in stable nuclides gradually increases as the number of protons in the nucleus increases.
- Light nuclides, such as ^{12}C, contain about the same number of neutrons and protons. Heavy nuclides, such as ^{238}U, contain up to 1.6 times as many neutrons as protons.
- There are no stable nuclides with atomic numbers larger than 83.
- This narrow band of stable nuclei is surrounded by a sea of instability.
- Nuclei that lie above this line have too many neutrons and are therefore neutron-rich.
- Nuclei that lie below this line don't have enough neutrons and are therefore neutron-poor.

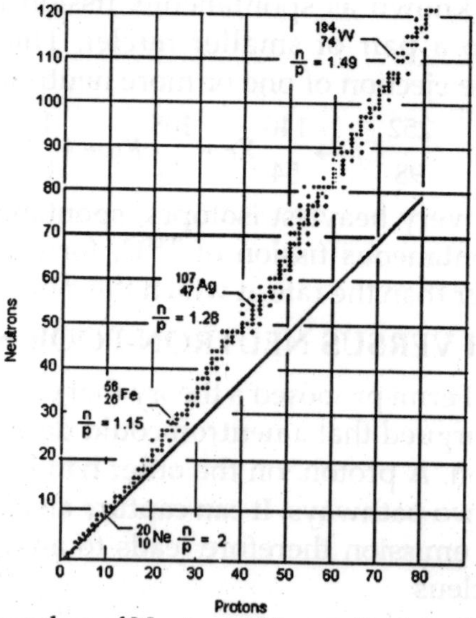

Fig. A Graph of the number of Neutrons Versus the Number of Protons for all Stable Naturally Occurring Nuclei.

The most likely mode of decay for a neutron-rich nucleus is one that converts a neutron into a proton. Every neutron-rich radioactive isotope with an atomic number smaller 83 decays by electron (β^-) emission. ^{14}C, ^{32}P, and ^{35}S, for example, are all neutron-rich nuclei that decay by the emission of an electron.

$$^{35}_{16}S \rightarrow ^{35}_{17}Cl + ^{0}_{-1}e \quad (\beta^- \text{emission})$$

Neutron-poor nuclides decay by modes that convert a proton into a neutron.

Neutron-poor nuclides with atomic numbers less than 83 tend to decay by either electron capture or positron emission. Many of these nuclides decay by both routes, but positron emission is more often observed in the lighter nuclides, such as ^{22}Na.

$$^{22}_{11}Na \rightarrow ^{22}_{10}Ne + ^{0}_{+1}e \quad (\beta^+ \text{emission})$$

Electron capture is more common among heavier nuclides, such as ^{125}I, because the 1s electrons are held closer to the nucleus of an atom as the charge on the nucleus increases.

$$^{125}_{53}I + ^{0}_{-1} \rightarrow ^{125}_{52}Te + h\nu \quad (\beta^+ \text{emission})$$

A third mode of decay is observed in neutron-poor nuclides that have atomic numbers larger than 83. Although it is not obvious at first, -decay increases the ratio of neutrons to protons. Consider what happens during the -decay of ^{238}U, for example.

$$^{238}_{92}U \rightarrow ^{234}_{90}Th + ^{4}_{2}He \quad (\alpha \text{ decay})$$

The parent nuclide (^{238}U) in this reaction has 92 protons and 146 neutrons, which means that the neutron-to-proton ratio is 1.587. The daughter nuclide (^{234}Th) has 90 protons and 144 neutrons, so its neutron-to-proton ratio is 1.600. The daughter nuclide is therefore slightly less likely to be neutron-pooow.

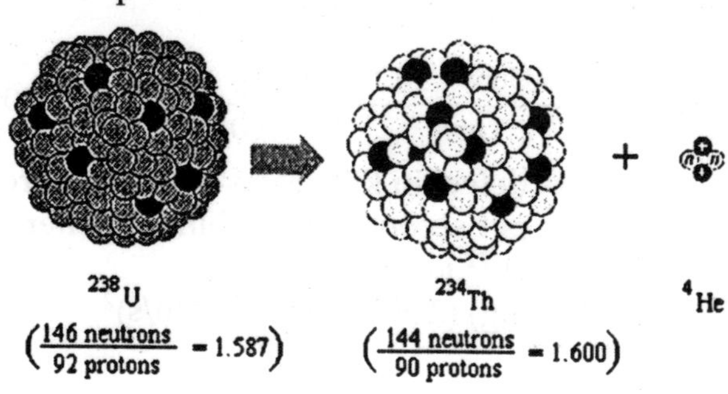

BINDING ENERGY CALCULATIONS

We should be able to predict the mass of an atom from the masses of the subatomic particles it contains. A helium atom, for example, contains two protons, two neutrons, and two electrons.

Example: The mass of a helium atom should be 4.0329802 amu.

Predicted mass = 4.0329802 amu
2(1.0072765) amu = 2.0145530 amu
2(1.0086650) amu = 2.0173300 amu
2(0.0005486) amu = 0.0010972 amu
Total mass = 4.0329802 amu

When the mass of a helium atom is measured, we find that the experimental value is smaller than the predicted mass by 0.0303769 amu.

Predicted mass = 4.0329802 amu
Observed mass = 4.0026033 amu
Mass defect = 0.0303769 amu

The difference between the mass of an atom and the sum of the masses of its protons, neutrons, and electrons is called the mass defect. The mass defect of an atom reflects the stability of the nucleus. It is equal to the energy released when the nucleus is formed from its protons and neutrons. The mass defect is therefore also known as the binding energy of the nucleus.

The binding energy serves the same function for nuclear reactions as H for a chemical reaction. It measures the difference between the stability of the products of the reaction and the starting materials. The larger the binding energy, the more stable the nucleus. The binding energy can also be viewed as the amount of energy it would take to rip the nucleus apart to form isolated neutrons and protons. It is therefore literally the energy that binds together the neutrons and protons in the nucleus.

The binding energy of a nuclide can be calculated from its mass defect with Einstein's equation that relates mass and energy.

$$E = mc^2$$

Example: We found the mass defect of He to be 0.0303769 amu. To obtain the binding energy in units of joules, we must convert the mass defect from atomic mass units to kilograms.

$$0.030796 amu \times \frac{1.6605655 \times 10^{-24}\, g}{1\, amu} \times \frac{1kg}{1000g} = 5.04428 \times 10^{-29}\, kg$$

Multiplying the mass defect in kilograms by the square of the speed of light in units of meters per second gives a binding energy for a single helium atom of 4.53358×10^{-12} joules.

$$E = (5.04428 \times 10^{-29} \text{ kg})\,(2.99792446 \times 10^{8} \text{ m/s})^2$$
$$= 4.53358 \times 10^{-12} \text{J}$$

Multiplying the result of this calculation by the number of atoms in a mole gives a binding energy for helium of 2.730 × 10^{12} joules per mole, or 2.730 billion kilojoules per mole.

$$\frac{4.5358\times10^{-12}\,J}{1\text{ atom}}\times\frac{6.022\times10^{23}\text{ atom}}{1\text{ mol}} = 2.730\times10^{12}\text{ J/mol}$$

This calculation helps us understand the fascination of nuclear reactions. The energy released when natural gas is burned is about 800 kJ/mol. The synthesis of a mole of helium releases 3.4 million times as much energy.

Since most nuclear reactions are carried out on very small samples of material, the mole is not a reasonable basis of measurement. Binding energies are usually expressed in units of electron volts (eV) or million electron volts (MeV) per atom.

Example: The binding energy of helium is 28.3 x 10^6 eV/atom or 28.3 MeV/atom.

$$\frac{4.53358\times10^{-12}\,J}{1\text{ atom}}\times\frac{1ev}{1.6021892\times10^{-19}\,J} = 28.30\times10^{6}\text{ ev/atom}$$

Calculations of the binding energy can be simplified by using the following conversion factor between the mass defect in atomic mass units and the binding energy in million electron volts.

1 amu = 931.5016 MeV

Binding energies gradually increase with atomic number, although they tend to level off near the end of the periodic table. A more useful quantity is obtained by dividing the binding energy for a nuclide by the total number of protons and neutrons it contains. This quantity is known as the binding energy per nucleon.

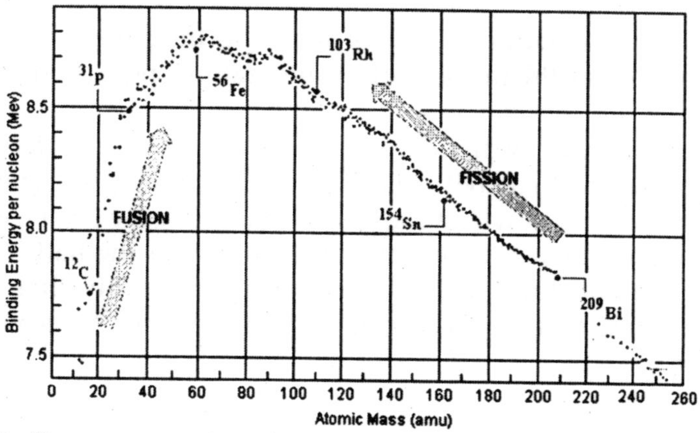

The binding energy per nucleon ranges from about 7.5 to 8.8 MeV for most nuclei. It reaches a maximum, however, at an atomic mass of about 60 amu.

The largest binding energy per nucleon is observed for ^{56}Fe, which is the most stable nuclide in the periodic table.

The graph of binding energy per nucleon versus atomic mass explains why energy is released when relatively small nuclei combine to form larger nuclei in fusion reactions.

$$^{12}_{6}C + ^{12}_{6}C \rightarrow ^{24}_{12}Mg \text{ (fussion)}$$

It also explains why energy is released when relatively heavy nuclei split apart in fission (literally, "to split or cleave") reactions.

$$^{235}_{92}U \rightarrow ^{139}_{56}Ba + ^{94}_{36}Kr + 2\,^{1}_{0}n \text{ (fussion)}$$

There are a number of small irregularities in the binding energy curve at the low end of the mass spectrum. The ^4He nucleus, for example, is much more stable than its nearest neighbors. The unusual stability of the ^4He nucleus explains why -particle decay is usually much faster than the spontaneous fission of a nuclide into two large fragments.

THE KINETICS OF RADIOACTIVE DECAY

Radioactive nuclei decay by first-order kinetics. The rate of radioactive decay is therefore the product of a rate constant (k) times the number of atoms of the isotope in the sample (N).

$$\text{Rate} = kN$$

The rate of radioactive decay doesn't depend on the chemical state of the isotope. The rate of decay of ^{238}U, for example, is exactly the same in uranium metal and uranium hexafluoride, or any other compound of this element.

The ra te at which a radioactive isotope decays is called the activity of the isotope. The most common unit of activity is the curie (Ci), which was originally defined as the number of disintegrations per second in 1 gram of ^{226}Ra.

The curie is now defined as the amount of radioactive isotope necessary to achieve an activity of 3.700×10^{10} disintegrations per second.

The relative rates at which radioactive nuclei decay can be expressed in terms of either the rate constants for the decay or the half-lives of the nuclei. We can conclude that ^{14}C decays more rapidly than ^{238}U, for example, by noting that the rate constant for the decay of ^{14}C is much larger than that for ^{238}U.

^{14}C: $k = 1.210 \times 10^{-4}\ y^{-1}$

^{238}U: $k = 1.54 \times 10^{-10}\ y^{-1}$

We can reach the same conclusion by noting that the half-life for the decay of ^{14}C is much shorter than that for ^{235}U.

^{14}C: $t_{1/2} = 5730$ y
^{238}U: $t_{1/2} = 4.51 \times 10^9$ y

The half-life for the decay of a radioactive nuclide is the length of time it takes for exactly half of the nuclei in the sample to decay.

In our discussion of the kinetics of chemical reactions, we concluded that the half-life of a first-order process is inversely proportional to the rate constant for this process.

$$t_{1/2} = \frac{\ln 2}{k} = \frac{0.693}{k}$$

The half-life of a nuclide can be used to estimate the amount of a radioactive isotope left after a given number of half-lives.

For more complex calculations, it is easier to convert the half-life of the nuclide into a rate constant and then use the integrated form of the first-order rate law described in the kinetic section.

DATING BY RADIOACTIVE DECAY

The earth is constantly bombarded by cosmic rays emitted by the sun. The total energy received in the form of cosmic rays is small no more than the energy received by the planet from starlight.

But the energy of a single cosmic ray is very large, on the order of several billion electron volts (200 million kJ/mol). These highly energetic rays react with atoms in the atmosphere to produce neutrons that then react with nitrogen atoms in the atmosphere to produce ^{14}C.

$$^{14}_{7}N + ^{1}_{0}n \rightarrow ^{14}_{6}C + ^{1}_{2}H$$

The ^{14}C formed in this reaction is a neutron-rich nuclide that decays by electron emission with a half-life of 5730 years.

$$^{14}_{6}C + ^{14}_{7}N + ^{1}_{-1}e$$

Just after World War II, Willard F. Libby proposed a way to use these reactions to estimate the age of carbon-containing substances. The ^{14}C dating technique for which Libby received the Nobel prize was based on the following assumptions.

- ^{14}C is produced in the atmosphere at a more or less constant rate.
- Carbon atoms circulate between the atmosphere, the oceans, and living organisms at a rate very much faster than they decay. As a result, there is a constant concentration of ^{14}C in all living things.
- After death, organisms no longer pick up ^{14}C.

Thus, by comparing the activity of a sample with the activity of living tissue we can estimate how long it has been since the organism *died*.

The natural abundance of ^{14}C is about 1 part in 10^{12} and the average activity of living tissue is 15.3 disintegrations per minute per gram of carbon. Samples used for ^{14}C dating can include charcoal, wood, cloth, paper, sea shells, limestone, flesh, hair, soil, peat, and bone. Since most iron samples also contain carbon, it is possible to estimate the time since iron was last fired by analyzing for ^{14}C.

We now know that one of Libby's assumptions is questionable: The amount of ^{14}C in the atmosphere hasn't been constant with time. Because of changes in solar activity and the earth's magnetic field, it has varied by as much as 5%.

More recently, contamination from the burning of fossil fuels and the testing of nuclear weapons has caused significant changes in the amount of radioactive carbon in the atmosphere. Radiocarbon dates are therefore reported in years before the present era (B.P.). By convention, the present era is assumed to begin in 1950, when ^{14}C dating was introduced.

Studies of bristlecone pines allow us to correct for changes in the abundance of ^{14}C with time. These remarkable trees, which grow in the White Mountains of California, can live for up to five thousand years. By studying the ^{14}C activity of samples taken from the annual growth rings in these trees, researchers have developed a calibration curve for ^{14}C dates from the present back to 5145 B.C.

After roughly 45,000 years (eight half-lives), a sample retains only 0.4% of the ^{14}C activity of living tissue. At that point it becomes too old to date by radiocarbon techniques. Other radioactive isotopes can be used to date rocks, soils, or archaeological objects that are much older. Potassium-argon dating, for example, has been used to date samples up to 4.3 billion years old.

Naturally occurring potassium contains 0.0118% by weight of the radioactive ^{40}K isotope. This isotope decays to ^{40}Ar with a half-life of 1.3 billion years. The ^{40}Ar produced after a rock crystallizes is trapped in the crystal lattice. It can be released, however, when the rock is melted at temperatures up to 2000C. By measuring the amount of ^{40}Ar released when the rock is melted and comparing it with the amount of potassium in the sample, the time since the rock crystallized can be determined.

IONIZING RADIATION

IONIZING VERSUS NON-IONIZING RADIATION

We live in a sea of radiation. In recent years, people have learned to fear the effects of radiation. They don't want to live near nuclear reactors. They are frightened by reports of links between excess exposure to sunlight and skin cancer. They are afraid of the leakage from microwave ovens, or the radiation produced by their television sets.

Several factors combine to heighten the public's anxiety about both the short-range and long-range effects of radiation. Perhaps the most important source of fear is the fact that radiation can't be detected by the average person. Furthermore, the effects of exposure to radiation might not appear for months or years or even decades.

To understand the biological effects of radiation we must first understand the difference between ionizing radiation and non-ionizing radiation.

In general, two things can happen when radiation is absorbed by matter: excitation or ionization.

- Excitation occurs when the radiation excites the motion of the atoms or molecules, or excites an electron from an occupied orbital into an empty, higher-energy orbital.
- Ionization occurs when the radiation carries enough energy to remove an electron from an atom or molecule.

Because living tissue is 70-90% water by weight, the dividing line between radiation that excites electrons and radiation that forms ions is often assumed to be equal to the ionization of water: 1216 kJ/mol.

Radiation that carries less energy can only excite the water molecule. It is therefore called non-ionizing radiation. Radiation that carries more energy than 1216 kJ/mol can remove an electron from a water molecule, and is therefore called ionizing radiation.

The energies of various kinds of radiation. Radio waves, microwaves, infrared radiation, and visible light are all forms of non-ionizing radiation. X-rays, -rays, and - and β -particles are forms of ionizing radiation.

The dividing line between ionizing and non-ionizing radiation in the electromagnetic spectrum falls in the ultraviolet portion of the spectrum.

It is therefore useful to divide the UV spectrum into two categories: UV_A and UV_B. Radiation at the high-energy end of the UV spectrum can be as dangerous as x-rays or -rays.

When ionizing radiation passes through living tissue, electrons are removed from neutral water molecules to produce H_2O^+ ions. Between three and four water molecules are ionized for every 1.6×10^{-17} joules of energy absorbed in the form of ionizing radiation.

$$H_2O \rightarrow H_2O^+ + e^-$$

The H_2O^+ ion should not be confused with the H_3O^+ ion produced when acids dissolve in water. The H_2O^+ ion is an example of a free radical, which contains an unpaired valence-shell electron. Free radicals are extremely reactive.

The radicals formed when ionizing radiation passes through water are among the strongest oxidizing agents that can exist in aqueous solution. At the molecular level, these oxidizing agents destroy biologically active molecules by either removing electrons or removing hydrogen atoms. This often leads to damage to the membrane, nucleus, chromosomes, or mitochondria of the cell that either inhibits cell division, results in cell death, or produces a malignant cell.

BIOLOGICAL EFFECTS OF IONIZING RADIATION

From the time that radioactivity was discovered, it was obvious that it caused damage. As early as 1901, Pierre Curie discovered that a sample of radium placed on his skin produced wounds that were very slow to heal. What some find surprising is the magnitude of the difference between the biological effects of non-ionizing radiation, such as light and microwaves, and ionizing radiation, such as high-energy ultraviolet radiation, x-rays, -rays, and - or β -particles.

Radiation at the low-energy end of the electromagnetic spectrum, such as radio waves and microwaves, excites the movement of atoms and molecules, which is equivalent to heating the sample. Radiation in or near the visible portion of the spectrum excites electrons into higher-energy orbitals. When the electron eventually falls back to a lower-energy state, the excess energy is given off to neighboring molecules in the form of heat. The principal effect of non-ionizing radiation is therefore an increase in the temperature of the system.

We experience the fact that biological systems are sensitive to heat each time we cook with a microwave oven, or spend too long in the sun. But it takes a great deal of non-ionizing radiation to reach dangerous levels. We can assume, for example, that absorption of enough radiation to produce an increase of about 6C in body temperature would be fatal. Since the average 70-kilogram human is 80% water by weight, we can use the heat capacity of water to calculate that it would take about 1.5 million joules of non-ionizing radiation to kill the average human. If this energy was carried by visible light with a frequency of 5×10^{14} s^{-1}, it would correspond to absorption of about seven moles of photons.

Ionizing radiation is much more dangerous. A dose of only 300 joules of x-ray or -ray radiation is fatal for the average human, even though this radiation raises the temperature of the body by only 0.001C. -particle radiation is even more dangerous; a dose equivalent to only 15 joules is fatal for the average human.

Whereas it takes seven moles of photons of visible light to produce a fatal dose of non-ionizing radiation, absorption of only 7×10^{-10} moles of the -particles emitted by ^{238}U is fatal.

There are three ways of measuring ionizing radiation.

- Measure the activity of the source in units of disintegrations per second or curies, which is the easiest measurement to make.
- Measure the radiation to which an object is exposed in units of roentgens by measuring the amount of ionization produced when this radiation passes through a sample of air.
- Measure the radiation absorbed by the object in units of radiation absorbed doses or "rads."
- This is the most useful quantity, but it is the hardest to obtain.

One radiation absorbed dose, or rad, corresponds to the absorption of 10^{-5} joules of energy per gram of body weight. Because this is equivalent to 0.01 J/kg, one rad produces an increase in body temperature of about 2×10^{-6}C.

At first glance, the rad may seem to be a negligibly small unit of measurement. The destructive power of the radicals produced when water is ionized is so large, however, that cells are inactivated at a dose of 100 rads, and a dose of 400 to 450 rads is fatal for the average human.

Not all forms of radiation have the same efficiency for damaging biological organisms. The faster energy is lost as the radiation passes through the tissue, the more damage it does.

To correct for the differences in radiation biological effectiveness (RBE) among various forms of radiation, a second unit of absorbed dose has been defined. The roentgen equivalent man, or rem, is the absorbed dose in rads times the biological effectiveness of the radiation.

$$\text{rems} = \text{rads} \times \text{RBE}$$

Values for the RBE of different forms of radiation are given.

Table. The Radiation Biological Effectiveness of Various Forms of Radiation.

Radiation	RBE
x-rays and -rays	1
β - particles with energies larger than 0.03 MeV	1
β - particles with energies less than 0.3 MeV	1.7
Thermal (slow-moving) neutrons	3
Fast-moving neutrons or protons	10
α-particles or heavy ions	20

Estimates of the per capita exposure to radiation. These estimates include both external and internal sources of natural background radiation.

Table. Average Whole-Body Exposure Levels for Sources of Ionizing Radiation

Source	Per Capita Dose (rems/y)
Natural background	0.082
Medical x-rays	0.077
Nuclear test fallout	0.005
Consumer and industrial products	0.005
Nuclear power industry	0.001
	total: 0.170

External sources include cosmic rays from the sun and -particles or -rays emitted from rocks and soil. Internal sources include nuclides that enter the body when we breathe (^{14}C, ^{85}Kr, ^{220}Rn, and ^{222}Rn) and through the food chain (^{3}H, ^{14}C, ^{40}K, ^{90}Sr, ^{131}I, and ^{137}Cs). The actual dose from natural radiation depends on where one lives. People who live in the Rocky Mountains, for example, receive twice as much background radiation as the national average because there is less atmosphere to filter out the cosmic rays from the sun.

The average dose from medical x-rays has decreased in recent years because of advances in the sensitivity of the photographic film used for x-rays. Radiation from nuclear test fallout has also decreased as a result of the atmospheric nuclear test ban. The threat of fallout from the testing of nuclear weapons can be appreciated by noting that a Chinese atmospheric test in 1976 led to the contamination of milk in the Harrisburg, PA, vicinity at a level of 300 pCi (3.00×10^{-10} Ci) per lite.

This was about eight times the level of contamination (41 pCi per liter) that resulted from the accident at Three Mile Island. The contribution to the radiation absorbed dose from consumer and industrial products includes radiation from construction materials, x-rays emitted by television sets, and inhaled tobacco smoke. The most recent estimate of the total radiation emitted from the mining and milling of uranium, the fabrication of reactor fuels, the storage of radioactive wastes, and the operation of nuclear reactors is less than 0.001 rem per year. The total dose from ionizing radiation for the average American is about 0.170 rem per year.

The Committee on the Biological Effects of Ionizing Radiation of the National Academy of Sciences recently estimated that an increase in this dose to a level of 1 rem per year would result in 169 additional deaths from cancer per million people exposed.

This can be compared with the 170,000 cancer deaths that would normally occur in a population this size that was not exposed to this level of radiation.

The principal effect of low doses of ionizing radiation is to induce cancers, which may take up to 20 years to develop. What is the effect of high doses of ionizing radiation? Cells that are actively dividing are more sensitive to radiation than cells that aren't.

Thus, cells in the liver, kidney, muscle, brain, and bone are more resistant to radiation than the cells of bone marrow, the reproductive organs, the epithelium of the intestine, and the skin, which suffer the most damage from radiation. Damage to the bone marrow is the main cause of death at moderately high levels of exposure (200 to 1000 rads).

Damage to the gastrointestinal tract is the major cause of death for exposures on the order of 100 to 10,000 rads. Massive damage to the central nervous system is the cause of death from extremely high exposures (over 10,000 rads).

NATURAL VERSUS INDUCED RADIOACTIVITY

NATURAL RADIOACTIVITY

The vast majority of the nuclides found in nature are stable. If our planet is 4.6 billion years old, the only radioactive isotopes that should remain are members of three classes.

- Isotopes with half-lives of at least 10^9 years, such as ^{238}U.
- Daughter nuclides produced when long-lived radioactive nuclides decay, such as the ^{234}Th ($t_{1/2}$ = 24.1 days) produced by the -decay of ^{238}U.
- Nuclides such as ^{14}C that are still being synthesized.

One factor that influences the stability of a nuclide is the ratio of neutrons to protons. (Nuclei that contain either too many or too few neutrons are unstable.) Another factor that affects the stability of nuclides can be understood by examining patterns in the numbers of protons and neutrons in stable nuclides.

Half of the elements in the periodic table must have an odd number of protons because atomic numbers that are odd are just as likely to occur as those that are even. In spite of this, about 80% of the stable nuclides have an even number of protons.

Very few elements with an odd atomic number have more than one stable isotope. Stable isotopes abound, however, among elements with even atomic numbers.

Ten stable isotopes are known for tin (Z = 50), for example. It is also interesting to note that 91% of the stable isotopes of elements with an odd number of protons have an even number of neutrons. These observations suggest that certain combinations of protons and neutrons are particularly stable.

There are magic numbers of electrons. Electron configurations with 2, 10, 18, 36, 54, and 86 electrons are unusually stable. There also seem to be magic numbers of neutrons and protons. Nuclei with 2, 8, 20, 28, 50, 82, or 126 protons or neutrons are unusually stable.

This observation explains the anomalously large binding energies observed for ^{4}He, ^{16}O, and ^{20}Ne. In each case, the nuclide has an even number of both protons and neutrons. ^{20}Ne has a magic number of nucleons when both protons and neutrons are counted. ^{4}He and ^{16}O have magic numbers of both protons and neutrons. The resulting stability of the ^{4}He nucleus might explain why so many heavy nuclei undergo -particle decay by ejecting an $^{4}He^{2+}$ ion from the nucleus of the atom.

If nuclei tend to be more stable when they have even numbers of protons and neutrons, it isn't surprising that nuclides with an odd number of both protons and neutrons are unstable. ^{40}K is one of only five naturally occurring nuclides that contain both an odd number of protons and an odd number of neutrons.

This nuclide simultaneously undergoes the electron capture and positron emission expected for neutron-poor nuclides and the electron emission observed with neutron-rich nuclides.

Only 18 radioactive isotopes with atomic numbers of 80 or less can be found in nature. With the exception of ^{14}C, which is continuously synthesized in the atmosphere, all these elements have lifetimes longer than 10^9 years. Although these isotopes all undergo radioactive decay, they decay so slowly that reasonable quantities are still present, 4.6 billion years after the planet was formed.

Another 45 natural radioactive isotopes have atomic numbers larger than 80. These nuclides fall into three families. The parent nuclide is ^{232}Th, which undergoes -decay to form ^{228}Ra. The product of this reaction decays by β -emission to form ^{228}Ac, which decays to ^{228}Th, and so on, until the stable ^{208}Pb isotope is formed. This family of radionuclides is called the 4n series, because all its members have a mass number that can be divided by 4.

A second family of radioactive nuclei starts with ^{238}U and decays to form the stable ^{206}Pb isotope. Every member of this series has a mass number that fits the equation 4n + 2. The third family, known as the 4n + 3 series, starts with ^{235}U and decays to ^{207}Pb. A 4n + 1 series once existed, which started with ^{237}Np and decayed to form the only stable isotope of bismuth, ^{209}Bi.

The half-life of every member of this series is less than 2×10^6 years, however, so none of the nuclides produced by the decay of neptunium remain in detectable quantities on the earth.

BINDING ENERGY

Binding energy is the mechanical energy required to disassemble a whole into separate parts. A bound system has typically a lower potential energy than its constituent parts; this is what keeps the system together. The usual convention is that this corresponds to a positive binding energy.

In general, binding energy represents the mechanical work which must be done in acting against the forces which hold an object together, while disassembling the object into component parts separated by sufficient distance that further separation requires negligible additional work. At the atomic level the atomic binding energy of the atom derives from electromagnetic interaction and is the energy required to disassemble an atom into free electrons and a nucleus. Electron binding energy is a measure of the energy required to free electrons from their atomic orbits.

At the nuclear level the nuclear binding energy (binding energy of nucleons into a nuclide) is derived from the strong nuclear force and is the energy required to disassemble a nucleus into the same number of free unbound neutrons and protons it is composed of, in such a way that the particles are far/distant enough from each other so that the strong nuclear force can no longer cause the particles to interact.

In astrophysics, gravitational binding energy of a celestial body is the energy required to expand the material to infinity. This quantity is not to be confused with the gravitational potential energy, which is the energy required to separate two bodies, such as a celestial body and a satellite, to infinite distance, keeping each intact (the latter energy is lower). In bound systems, if the binding energy is removed from the system, it must be subtracted from the mass of the unbound system, simply because this energy has mass, and if subtracted from the system at the time it is bound, will result in removal of mass from the system. System mass is not conserved in this process because the system is not closed during the binding process

Mass Deficit

Classically a bound system is at a lower energy level than its unbound constituents, its mass must be less than the total mass of its unbound constituents. For systems with low binding energies, this "lost" mass after binding may be fractionally small. For systems with high binding energies, however, the missing mass may be an easily measurable fraction.

Since all forms of energy have mass, the question of where the missing mass of the binding energy goes, is of interest. The answer is that this mass is lost from a system which is not closed. It transforms to heat, light, higher energy states of the nucleus/atom or other forms of energy, but these types of energy also have mass, and it is necessary that they be removed from the system before its mass may decrease. The "mass deficit" from binding energy is therefore removed mass that corresponds with removed energy, according to Einstein's equation $E = mc^2$. Once the system cools to normal temperatures and returns to ground states in terms of energy levels, there is less mass remaining in the system than there was when it first combined and was at high energy. Mass measurements are almost always made at low temperatures with systems in ground states, and this difference between the mass of a system and the sum of the masses of its isolated parts is called a mass deficit. Thus, if binding energy mass is transformed into heat, the system must be cooled (the heat removed) before the mass-deficit appears in the cooled system. In that case, the removed heat represents exactly the mass "deficit", and the heat itself retains the mass which was lost.

The attraction force accelerates the objects and they gain some speed toward each other converting the potential (gravity) energy into kinetic (movement) energy. When either the particles 1) pass through each other without interaction or 2) elastically repel during the collision, the gained kinetic energy (related to speed), starts to revert into potential form driving the collided particles apart. The decelerating particles will return to the initial distance and beyond into infinity or stop and repeat the collision (oscillation takes place). This shows that the system, which loses no energy, does not combine (bind) into a solid object, parts of which oscillate at short distances.

Therefore, in order to bind the particles, the kinetic energy gained due to the attraction must be dissipated (by resistive force). Complex objects in collision ordinarily undergo inelastic collision, transforming some kinetic energy into internal energy (heat content, which is atomic movement), which is further radiated in the form of photons-the light and heat. Once the energy to escape the gravity is dissipated in the collision, the parts will oscillate at closer, possibly atomic, distance, thus looking like one solid object. This lost energy, necessary to overcome the potential barrier in order to separate the objects, is the binding energy. If this binding energy were retained in the system as heat, its mass would not decrease. However, binding energy lost from the system (as heat radiation) would itself have mass, and directly represent of the "mass deficit" of the cold, bound system.

Closely analogous considerations apply in chemical and nuclear considerations. Exothermic chemical reactions in closed systems do not change mass, but (in theory) become less massive once the heat of reaction is removed. This mass change is too small to measure with standard equipment. In nuclear reactions, however, the fraction of mass that may be removed as light or heat, i.e., binding energy, is often a much larger fraction of the system mass. It may thus be measured directly as a mass difference between rest masses of reactants and products. This is because nuclear forces are comparatively stronger than Coulombic forces associated with the interactions between electrons and protons that generate heat in chemistry.

Mass Defect

In simple words definition of mass defect can be stated as follows:

Definition: The distance between theoretical calculated mass and experimentally measured mass of nucleus is called mass defect. It is denoted by Δm. It can be calculated as follows:

Mass defect = (Theoretical calculated mass) – (measured mass of nucleus)

i.e, (sum of masses of protons and neutrons) – (measured mass of nucleus)

In nuclear reactions, the energy that must be radiated or otherwise removed as binding energy may be in the form of electromagnetic waves, such as gamma radiation, or as heat. Again, however, no mass deficit can in theory appear until this radiation has been emitted and is no longer part of the system. The energy given off during either nuclear fusion or nuclear fission is the difference between the binding energies of the fuel and the fusion or fission products.

In practice, this energy may also be calculated from the substantial mass differences between the fuel and products, once evolved heat and radiation have been removed. When the nucleons are grouped together to form a nucleus, they lose a small amount of mass, i.e., there is mass defect. This mass defect is released as (often radiant) energy according to the relation $E = mc^2$; thus binding energy = mass defect $\cdot c^2$. This energy holds the nucleons together and is known as binding energy. In fact, mass defect is a measure of the binding energy of the nucleus. The greater the mass defect, the greater is the binding energy of the nucleus.

Mass Excess

It is observed experimentally that the mass of the nucleus is smaller than the number of nucleons each counted with a mass of 1 a.m.u.. This difference is called mass excess. The difference between the actual mass of the nucleus measured in atomic mass units and the number of nucleons is called mass excess i.e. with M equals the actual mass of the nucleus in u and A equals the mass number.

$$\text{Mass excess} = M - A = \text{Excess-energy} / c^2$$

with M equals the actual mass of the nucleus in u and A equals the mass number.

NUCLEAR FISSION AND NUCLEAR FUSION

NUCLEAR FISSION

The graph of binding energy per nucleon suggests that nuclides with a mass larger than about 130 amu should spontaneously split apart to form lighter, more stable, nuclides. Experimentally, we find that spontaneous fission reactions occur for only the very heaviest nuclides those with mass numbers of 230 or more.

Even when they do occur, these reactions are often very slow. The half-life for the spontaneous fission of ^{238}U, for example, is 10^{16} years, or about two million times longer than the age of our planet!

We don't have to wait, however, for slow spontaneous fission reactions to occur. By irradiating samples of heavy nuclides with slow-moving thermal neutrons it is possible to induce fission reactions. When ^{235}U absorbs a thermal neutron, for example, it splits into two particles of uneven mass and releases an average of 2.5 neutrons.

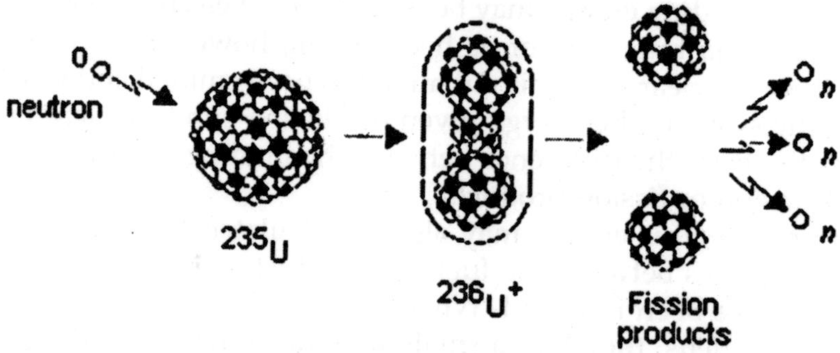

Fig. The Absorption of a Neutron by ^{238}U induces Oscillations in the Nucleus that deform it until it splits into Fragments the way a Drop of Liquid might break into smaller Droplets.

More than 370 daughter nuclides with atomic masses between 72 and 161 amu are formed in the thermal-neutron-induced fission of ^{235}U, including the two products shown below.

$$^{235}_{92}U + ^{1}_{0}n \rightarrow ^{139}_{56}Ba + ^{94}_{36}Kr + 3\,^{1}_{0}n$$

Several isotopes of uranium undergo induced fission. But the only naturally occurring isotope in which we can induce fission with thermal neutrons is ^{235}U, which is present at an abundance of only 0.72%. The induced fission of this isotope releases an average of 200 MeV per atom, or 80 million kilojoules per gram of ^{235}U. The attraction of nuclear fission as a source of power can be understood by comparing this value with the 50 kJ/g released when natural gas is burned.

The first artificial nuclear reactor was built by Enrico Fermi and co-workers beneath the University of Chicago's football stadium and brought on line on December 2, 1942. This reactor, which produced several kilowatts of power, consisted of a pile of graphite blocks weighing 385 tons stacked in layers around a cubical array of 40 tons of uranium metal and uranium oxide.

Spontaneous fission of ^{238}U or ^{235}U in this reactor produced a very small number of neutrons. But enough uranium was present so that one of these neutrons induced the fission of a ^{235}U nucleus, thereby releasing an average of 2.5 neutrons, which catalyzed the fission of additional ^{235}U nuclei in a chain reaction. The amount of fissionable material necessary for the chain reaction to sustain itself is called the critical mass.

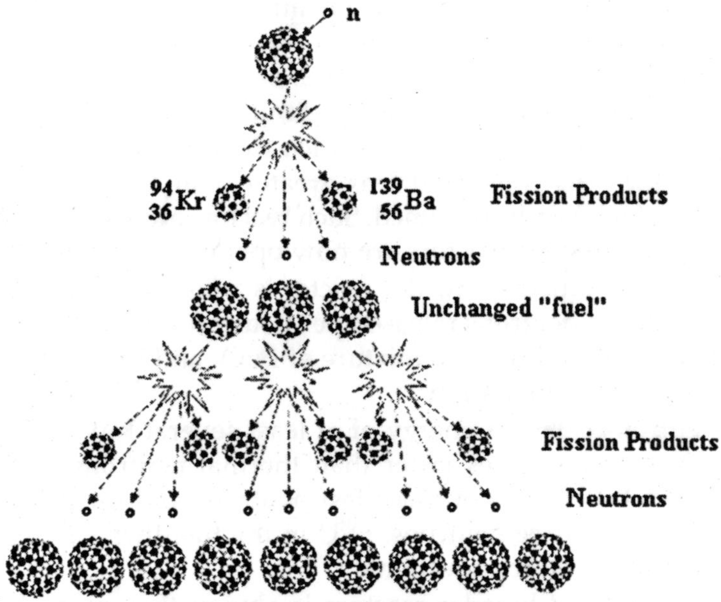

Fig. Fission Reaction.

The Fermi reactor at Chicago served as a prototype for larger reactors constructed in 1943 at Oak Ridge, Tennessee, and Hanford, Washington, to produce ^{239}Pu for one of the atomic bombs dropped on Japan at the end of World War II. As we have seen, some of the neutrons released in the chain reaction are absorbed by ^{238}U to form ^{239}U, which undergoes decay by the successive loss of two β -particles to form ^{239}Pu. ^{238}U is an example of a fertile nuclide. It doesn't undergo fission with thermal neutrons, but it can be converted to ^{239}Pu, which does undergo thermal-neutron-induced fission.

Fission reactors can be designed to handle naturally abundant ^{235}U, as well as fuels described as slightly enriched (2-5% ^{235}U), highly enriched

(20-30% ^{235}U), or fully enriched (>90% ^{235}U). Heat generated in the reactor core is transferred to a cooling agent in a closed system. The cooling agent is then passed through a series of heat exchangers in which water is heated to steam.

The steam produced in these exchangers then drives a turbine that generates electrical power. There are two ways of specifying the power of such a plant: the thermal energy produced by the reactor or the electrical energy generated by the turbines. The electrical capacity of the plant is usually about one-third of the thermal power.

It takes 10^{11} fissions per second to produce one watt of electrical power. As a result, about one gram of fuel is consumed per day per megawatt of electrical energy produced. This means that one gram of waste products is produced per megawatt per day, which includes 0.5 grams of ^{239}Pu. These waste products must be either reprocessed to generate more fuel or stored for the tens of thousands of years it takes for the level of radiation to reach a safe limit.

If we could design a reactor in which the ratio of the ^{239}Pu or ^{233}U produced to the ^{235}U consumed was greater than 1, the reactor would generate more fuel than it consumed. Such reactors are known as breeders, and commercial breeder reactors are now operating in France.

The key to an efficient breeder reactor is a fuel that gives the largest possible number of neutrons released per neutron absorbed. The breeder reactors being built today use a mixture of PuO_2 and UO_2 as the fuel and fast neutrons to activate fission.

Fast neutrons carry an energy of at least several KeV and therefore travel 10,000 or more times faster than thermal neutrons. ^{239}Pu in the fuel assembly absorbs one of these fast neutrons and undergoes fission with the release of three neutrons. ^{238}U in the fuel then captures one of these neutrons to produce additional ^{239}Pu.

The advantage of breeder reactors is obviousthey mean a limitless supply of fuel for nuclear reactors. There are significant disadvantages, however. Breeder reactors are more expensive to build. They are also useless without a subsidiary industry to collect the fuel, process it, and ship the ^{239}Pu to new reactors.

It is the reprocessing of ^{239}Pu that concerns most of the critics of breeder reactors. ^{239}Pu is so dangerous as a carcinogen that the nuclear industry places a limit on exposure to this material that assumes workers inhale no more than 0.2 micrograms of plutonium over their lifetimes. There is also concern that the ^{239}Pu produced by these reactors might be stolen and assembled into bombs by terrorist organizations.

The fate of breeder reactors in the United States is linked to economic considerations. Because of the costs of building these reactors and safely reprocessing the ^{239}Pu produced, the breeder reactor becomes economical only when the scarcity of uranium drives its price so high that the breeder reactor becomes cost effective by comparison. If nuclear energy is to play a dominant role in the generation of electrical energy in the 21st century, breeder reactors eventually may be essential.

Although the "pile" Fermi constructed at the University of Chicago in 1942 was the first artificial nuclear reactor, it was not the first fission reactor to exist on Earth.

In 1972, a group of French scientists discovered that uranium ore from a deposit in the Oklo mine in Gabon, West Africa, contained 0.4% ^{235}U instead of the 0.72% abundance found in all other sources of this ore.

Analysis of the trace elements in the ore suggested that the amount of ^{235}U in this ore was unusually small because natural fission reactors operated in this deposit for a period of 600,000 to 800,000 years about 2 billion years ago.

NUCLEAR FUSION

The graph of binding energy per nucleon suggests another way of obtaining useful energy from nuclear reactions. Fusing two light nuclei can liberate as much energy as the fission of ^{235}U or ^{239}Pu. The fusion of four protons to form a helium nucleus, two positrons (and two neutrinos), for example, generates 24.7 MeV of energy.

$$4\,{}^{1}_{1}H \rightarrow {}^{4}_{2}He + 2\,{}^{0}_{+1}e$$

Most of the energy radiated from the surface of the sun is produced by the fusion of protons to form helium atoms within its core.

Fusion reactions have been duplicated in man-made devices. The enormous destructive power of the ^{235}U-fueled atomic bomb dropped on Hiroshima on August 6, 1945, which killed 75,000 people, and the ^{239}Pu-fueled bomb dropped on Nagasaki three days later touched off a violent debate after World War II about the building of the next superweapon a fusion, or "hydrogen," bomb. Alumni of the Manhattan project, who had developed the atomic bomb, were divided on the issue.

Ernest Lawrence and Edward Teller fought for the construction of the fusion device. J. Robert Oppenheimer and Enrico Fermi argued against it. The decision was made to develop the weapon, and the first artificial fusion reaction occurred when the hydrogen bomb was tested in November 1952.

The history of fusion research is therefore the opposite of fission research. With fission, the reactor came first, and then the bomb was built.

With fusion, the bomb was built long before any progress was made toward the construction of a controlled fusion reactor. More than 40 years after the first hydrogen bomb was exploded, the feasibility of controlled fusion reactions is still open to debate.

The reaction that is most likely to fuel the first fusion reactor is the thermonuclear d-t, or deuterium-tritium, reaction. This reaction fuses two isotopes of hydrogen, deuterium (^2H) and tritium (^3H), to form helium and a neutron.

$$ {}^{2}_{1}H + {}^{3}_{1}H \rightarrow {}^{2}_{2}He + {}^{1}_{0}n $$

If we consider the implications of this reaction we can begin to understand why it is called a thermonuclear reaction and why it is so difficult to produce in a controlled manner. The d-t reaction requires that we fuse two positively charged particles.

This means that we must provide enough energy to overcome the force of repulsion between these particles before fusion can occur. To produce a self-sustaining reaction, we have to provide the particles with enough thermal energy so that they can fuse when they collide.

Each fusion reaction is characterized by a specific ignition temperature, which must be surpassed before the reaction can occur. The d-t reaction has an ignition temperature above 10^8 K. In a hydrogen bomb, a fission reaction produced by a small atomic bomb is used to heat the contents to the temperature required to initiate fusion. Obtaining the same result in a controlled reaction is much more challenging.

Any substance at temperatures approaching 10^8 K will exist as a completely ionized gas, or plasma. The goals of fusion research at present include the following.

- To achieve the required temperature to ignite the fusion reaction.
- To keep the plasma together at this temperature long enough to get useful amounts of energy out of the thermonuclear fusion reactions.
- To obtain more energy from the thermonuclear reactions than is used to heat the plasma to the ignition temperature.

These are not trivial goals. The only reasonable container for a plasma at 10^8 K is a magnetic field. Both doughnut-shaped (toroidal) and linear magnetic bottles have been proposed as fusion reactors. But reactors that produce high enough temperatures for ignition are not the same as the reactors that have produced long enough confinement times for the plasma to provide useful amounts of energy.

A second approach to a controlled fusion reactor involves hitting fuel pellets containing the proper reagents for the thermonuclear reaction with pulsed beams of laser power. If enough power was delivered, the fuel pellets would collapse upon themselves, or implode, to reach densities several orders of magnitude greater than normal. This could produce a plasma both hot enough and dense enough to initiate fusion reactions.

NUCLEAR SYNTHESIS AND NUCLEAR MEDICINE

NUCLEAR SYNTHESIS

Within the last 60 years, scientists have generated a new story of creation that offers a model for understanding how nuclei are synthesized. In 1929, Edwin Hubble provided evidence to suggest that our universe is expanding. Between 1946 and 1948, George Gamow and co-workers generated a model which assumed that the primordial substance, or ylem, from which all other matter was created was an extraordinarily hot, dense singularity that exploded in a "Big Bang" and has been expanding ever since.

This model assumes that neutrons in the ylem were transformed into protons by β -decay. Neutrons and protons then combined to form ^4He atoms before the temperature and pressure of the fireball decayed to the point at which no further nuclear reactions were possible.

The first-generation stars condensed out of this cloud of hydrogen and helium. As the gas condensed by gravitational attraction, it became warmer. Eventually, the temperature reached 10^7 K, and first-generation, main-sequence stars were born. The temperature at the cores of these stars was high enough to ignite the following thermonuclear reactions.

$$ {}^{1}_{1}H + {}^{1}_{1}H \rightarrow {}^{2}_{1}H + {}^{1}_{0}e $$

$$ {}^{1}_{1}H + {}^{2}_{1}H \rightarrow {}^{3}_{2}He $$

$$ {}^{3}_{2}He + {}^{3}_{2}He \rightarrow {}^{4}_{2}He + 2\,{}^{1}_{1}H $$

The net result of these reactions is the formation of a helium atom from four protons.

$$ 4\,{}^{1}_{1}H \rightarrow {}^{4}_{2}He + {}^{0}_{+1}e $$

Eventually the heat generated in this reaction was enough to halt the gravitational collapse of the star, which entered a stable period during which the energy generated by this reaction balanced the energy radiated at the surface.

The hydrogen-burning reactions in a main-sequence star are concentrated in the core. When enough hydrogen has been consumed, the core begins to collapse, and the temperature of the core rises above 10^8 K. (The larger the star, the more rapidly it radiates energy from its surface, and the more rapidly it consumes the hydrogen in the core).

As the core collapses, the hydrogen-containing outer shell expands, and the surface of the star cools. Stars that have reached this point in their evolution include the so-called red giants.

Temperatures in the core of these red giants are high enough to ignite further thermonuclear fusion reactions, such as the following.

$$3\,{}^{4}_{2}He \rightarrow {}^{12}_{6}C$$

This reaction becomes the principal source of energy in a red giant, although there is undoubtedly some burning of hydrogen to helium in the outer shell of the star. As the amount of ^{12}C in the core increases, further reactions occur to form ^{16}O and ^{20}Ne.

$${}^{12}_{6}C + {}^{4}_{2} \rightarrow {}^{16}_{8}O + \gamma$$

$${}^{16}_{8}O + {}^{4}_{2}He \rightarrow {}^{20}_{10}Ne + \gamma$$

Eventually, the helium in the core is exhausted, and the core collapses further, reaching temperatures of 6-7 x 10^8 K. At this point, more complex reactions take place that produce nuclides such as ^{28}Si and ^{32}S.

$${}^{12}_{6}C + {}^{16}_{8}O \rightarrow {}^{28}_{14}Si + \gamma$$

$${}^{16}_{8}O + {}^{16}_{8}O \rightarrow {}^{32}_{16}S + \gamma$$

Further gravitational collapse heats the core to temperatures above 10^9 K, and a complex sequence of reactions takes place to synthesize the nuclei with the highest binding energies, such as Fe and Ni.

If the star explodes in a supernova, its contents are ejected across space. Second-generation stars that condense in this region contain not only hydrogen and helium but elements with higher atomic number.

The best estimates of the age of the Milky Way suggest that our galaxy is about 15 billion years old. Our sun and its planets, however, are only 4.6 billion years old. This suggests that the sun is a second-generation star. In such stars, the transformation of hydrogen to helium can be catalyzed by ^{12}C.

$$^{12}_{6}C + ^{1}_{1}H \rightarrow ^{13}_{7}N + \gamma$$

$$^{13}_{7}N \rightarrow ^{13}_{6}C + ^{0}_{+1}$$

$$^{13}_{6}C + ^{1}_{1}H \rightarrow ^{14}_{7}N + \gamma$$

$$^{14}_{6}N + ^{1}_{1}H \rightarrow ^{15}_{8}N + \gamma$$

$$^{18}_{8}O \rightarrow ^{15}_{7}N + ^{0}_{+1}e$$

$$^{15}_{7}N + ^{1}_{1}H \rightarrow ^{12}_{6}C + ^{4}_{2}He$$

Second-generation stars use a different mechanism to synthesis helium than first-generation stars. They use a sequence of reactions, such as the those shown here, that are catalyzed by isotopes heavier than helium.

Two processes can synthesize elements with atomic numbers larger than that of iron. One of them is relatively slow (s-process), the other is very rapid (r-process). Since the only way to synthesize nuclei with atomic numbers larger than iron is by the absorption of neutrons, both the s-process and the r-process result from (n,) reactions.

In the s-process, neutrons are captured one at a time to form a neutron-rich nuclide that has enough time to undergo - or β -decay before another neutron can be absorbed. An example of an s-process sequence of reactions starts with ^{120}Sn. The capture of a neutron produces ^{121}Sn, which undergoes β^--decay.

If β^--decay occurs before this nuclide captures another neutron, a stable isotope of antimony is formed. Eventually, ^{121}Sb captures a neutron to produce ^{122}Sb, which is transformed into ^{122}Te by β -decay. ^{122}Te can undergo β -decay to form ^{122}I, or it can capture a neutron to form ^{123}Te. With ^{123}Te, we encounter a series of stable isotopes of tellurium. Thus, neutrons are slowly absorbed, one at a time, until we reach ^{127}Te, which decays to ^{127}I, the most abundant isotope of iodine.

$$^{120}_{60}Sn(n,\gamma)\,^{121}_{50}Sn \overset{\beta^-}{\rightarrow} ^{121}_{51}Sb(n,\gamma)$$

$$^{122}_{51}Sb \overset{\beta^-}{\rightarrow} ^{122}_{52}Te(5n,5y)\,^{127}_{52}Te \overset{\beta^-}{\rightarrow} ^{127}_{53}I$$

This slow process can't account for very heavy nuclides, such as ^{232}Th and ^{238}U, because the lifetimes of the intermediate nuclei with atomic numbers between 83 and 90 are too short for this step-by-step absorption of neutrons to proceed.

Synthesizing appreciable quantities of uranium and thorium requires a rapid process. In the r-process, a number of neutrons are captured in rapid succession, before there is time for - or β -decay to take place.

Achieving an r-process reaction, however, requires a very high neutron flux. (These reactions occur during nuclear explosions, for example.) The neutron flux needed to fuel such reactions is not likely to occur in a normal star.

During the moment when a star explodes as a supernova, however, the conditions are ripe for r-process reactions. The heavier elements on this planet were therefore produced in a series of supernova explosions that occurred in this portion of the galaxy before our solar system condensed.

NUCLEAR MEDICINE

Ever since the first x-ray images were obtained by Roentgen in 1895, ionizing radiation and radio nuclides have played a vital role in medicine. This work has been so fruitful that a separate field known as nuclear medicine has developed. Research in this field focuses on either therapeutic or diagnostic uses of radiation.

There are three standard approaches to fighting cancer: surgery, chemotherapy, and radiation. Surgery, by its very nature, is invasive. Chemotherapy and classic approaches to radiation therapy are not selective. Research in recent years has therefore examined new approaches to radiation therapy that specifically attack tumor cells, without damaging normal tissue. The technique known as boron neutron capture therapy provides an example of this work.

Naturally occurring boron consists of two stable isotopes: ^{10}B (19.7%) and ^{11}B (80.3%). ^{10}B absorbs thermal neutrons to form ^{11}B in a nuclear excited state. Although ^{11}B in its nuclear ground state is stable, this excited ^{11}B nuclide undergoes fission to produce ^{7}Li and an α-particle.

$$^{10}_{5}B + ^{1}_{0}n \rightarrow ^{7}_{3}Li + ^{4}_{2}He$$

Because the energy of a thermal neutron is only about 0.025 eV, the neutrons that are not absorbed do relatively little damage to the normal tissue. The -particle emitted in this reaction has an energy of 2.79 MeV, however, which makes it an extremely lethal form of radiation.

The RBE for -particle radiation is larger than most other particles because this relatively massive particle loses energy very efficiently as it collides with matter. Radiation damage from the -particle is therefore restricted to the immediate vicinity of the tissue that absorbed the thermal neutron.

Although other common nuclides in living tissue can absorb thermal neutrons, the ability of ^{10}B to absorb thermal neutrons is three orders of magnitude larger than these nuclides. The units with which this measurement is made can be understood by thinking about the area around the nuclide through which the neutron can pass and still be absorbed.

The neutron-capture cross section for a hydrogen atom corresponds to a circle with a radius of about 2×10^{-13} cm. For a nitrogen atom, the radius of this circle is about 10^{-12} cm. Boron has a neutron-capture cross section that would be described by a circle with a radius of about 2×10^{-9} cm. Virtually all of the neutron capture that occurs is therefore concentrated in the tissue that contains boron.

The potential of boron neutron capture therapy was recognized as early as 1936. Early clinical trials in the late 1950s and early 1960s failed to prolong the lives of patients suffering from brain tumors because the boron compounds available for testing did not concentrate selectively in the tumor cells. Recent research has discovered boron-labeled compounds that are sufficiently selective for BNCT to become useful as an adjunct to surgery, or in place of surgery with inoperable cases.

Chapter 7

Electrostatic Force

Electrostatics is associated with materials on which electrical charge moves only slowly (insulating materials) and with electrically isolated conductors. 'Insulation' and 'isolation' prevents easy migration of charge. So charges stay in place (are 'static'). It is the effects that the charges produce that are then important. The parameter of basic importance in electrostatics is 'charge'.

It is charge which gives rise to the electric fields generating forces that attract thin films and particles to surfaces and charge which gives rise to high voltages in low capacitance systems. High voltages can cause sparks. If these have sufficient energy they may ignite flammable gases, cause shocks to personnel, damage semiconductor devices and upset the operation of microelectronic equipment.

CHARGE

Static electricity arises as the separation of positive and negative charges at the interface between two dissimilar surfaces. If one or other of the surfaces prevent easy migration of charge, or the conductor on which they reside is isolated, then this charge is 'static' on the surface and remains available to influence the surroundings. 'Static' electricity can also arise on surfaces as trapped ions from the air.

Static charges may be electrons, or positive, or negative ions–but they are in the basic units of electronic charge 1.602×10^{-19} coulomb. On a surface there are some 10^{19} atomic sites per m^{-2}so if the charge of even quite a small fraction of the surface atoms is changed then quite large quantities of charge are easily involved.

FORCE BETWEEN CHARGES

The force between two charges, Q_1 and Q_2, is proportional to the quantities of charge, and inversely proportional to the square of the separation distance, d:

$$F = -Q_1\, Q_2/(4\pi\varepsilon_0 d^2)$$

This is Coulomb's Law. The force F is in Newtons, the quantities of charge Q_1 and Q_2 are in Coulombs and the separation distance d is in meters. The constant ε_0 is called the 'permittivity of free space' and has a value 8.854×10^{-12}. Charges of the same polarity repel each other and of opposite polarity attract.

Examples of electrostatic forces are the attraction between thin layers of plastic film and of fine dust particles to surfaces.

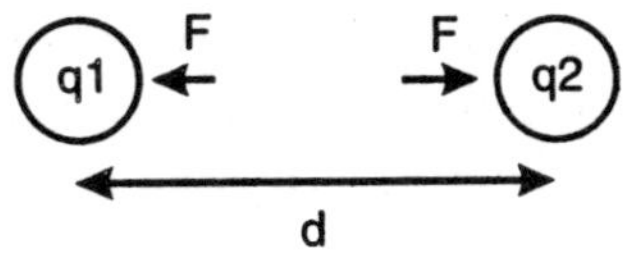

ELECTRIC FIELD

Around a charged body there is a force of attraction or repulsion for any other charges. This force, for a unit of charge, at any point is called the 'electric field', E, at that point. The direction of the field at any point depends on the direction of force on a positive charge there–and is hence a vector parameter. Diagrammatically it is useful to draw 'lines of force' to represent the electric field.

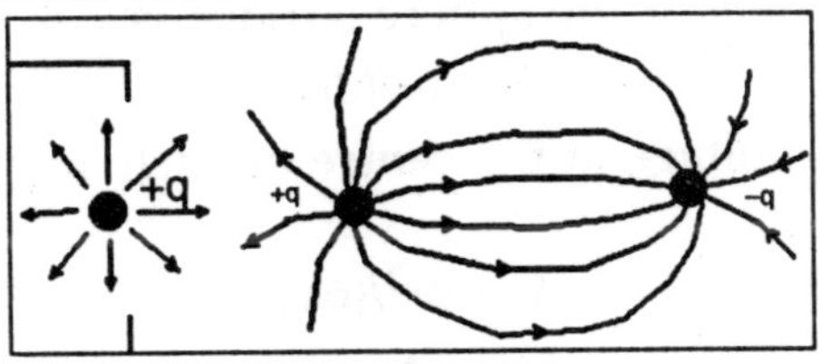

Fig. The Lines of Force the Stronger the Electric Field.

From Coulombs Law the electric field strength E at a distance d from a charge Q will be:

$$E = Q/(4\pi\varepsilon_0 d^2)$$

The constant of proportionality $1/(4\pi\varepsilon_0)$ is, for historical reasons, defined in the MKS system of units by the relation:

$$1/(4\pi\varepsilon_0) = c^2 \times 10^{-7}$$

The constant ε_0 hence has a value 8.854×10^{-12} coulomb^2/(Newton m^2) or coulomb/(volt m). This is the 'permittivity of free space' noted above.

POTENTIAL

The 'potential' at a point is defined as the amount of work needed to bring a unit charge from infinity to that point. The potential difference between two points is then the work done to move a unit charge between these two points. The work done does not depend on the route followed so the potential is a scalar quantity.

The unit of potential is the 'volt'. The potential at a radial distance d from a single point charge Q is:

$$V = Q/(4\pi\varepsilon_0 d)$$

The electric field relates to potential as:

$$E = -\nabla V$$

ELECTRIC FLUX }DENSITY

The electric field from an isolated point charge is radial. The field may conveniently be represented diagrammatically by lines of force. The 'flow' or flux of electric field across any spherical surface enclosing a point charge Q will then be Q/ε_o.

GAUSS' LAW

Gauss' Law states that the nett flux through any surface equals the nett charge enclosed within that surface.

$$\nabla.E = q/\varepsilon_0$$

For a uniform distribution density of charge ρ (C m^{-3}) in a sphere of radius r the total charge is $4/3\ \pi\ r^3\ \rho$. The outward flux will be $4\ \pi\ r^2\ E$. Hence the electric field at the sphere surface is:

$$E = \rho r/(3\varepsilon_0)$$

The maximum voltage at the centre of such a spherical distribution of charge is:

$$V = \rho r^2/(6\varepsilon_0)$$

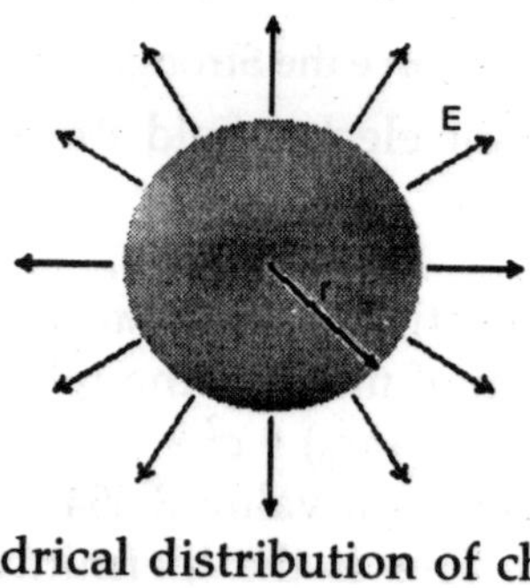

Similarly for a cylindrical distribution of charge density:

$$E = \rho r/(2\varepsilon_0) \text{ and } V = \rho r^2/(4\varepsilon_0)$$

INDUCED CHARGE

An electric field will have charge 'induced' on it with the density of charge proportional to the electric field. The charge induced on a conductor is of opposite polarity to that of the charge source. Opposite charges attract so there is an attraction between a charge and a conducting surface.

For a plane conducting surface this 'image charge' is an equal distance behind the surface so calculation of the force involves a distance equal to twice the separation distance from the conducting surface.

The electric field from each side of a sheet of uniform charge density σ (C m^{-2}) is:

$$E = \sigma/(2\varepsilon_0)$$

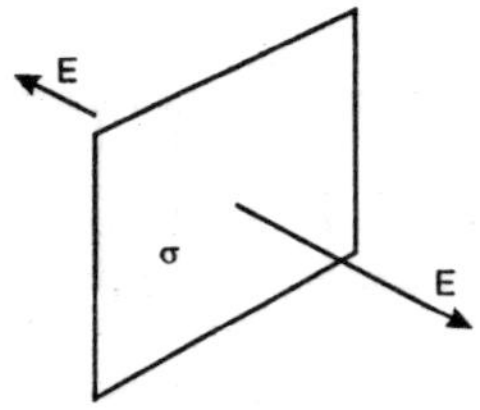

Fig. Electric Field from each side of a Sheet of Uniform Charge density

There can be no electric fields inside a conductor and the electric field outside is:

$$E = \sigma/\varepsilon_0$$

If a small conducting body (e.g. a small sphere) on an insulating handle is touched to a conducting body charge will be shared and the probe will acquire a charge dependent of the local electric field–related to the local surface charge density.

If the amount of charge can be tested then we have a means to check if the conducting body was charged and to explore the density of charge over that body.

A 'gold leaf electroscope' indicates charge by repulsion between a fine gold leaf from it's nearby conducting support and gives an angle of the leaf indicating the level of charge. Sharing charge from a conducting probe with a gold leaf electroscope provides a good classical method to indicate and measure charge.

Such observations show that the electric field (and charge density) on a conducting body is inversely proportional to the local radius of curvature.

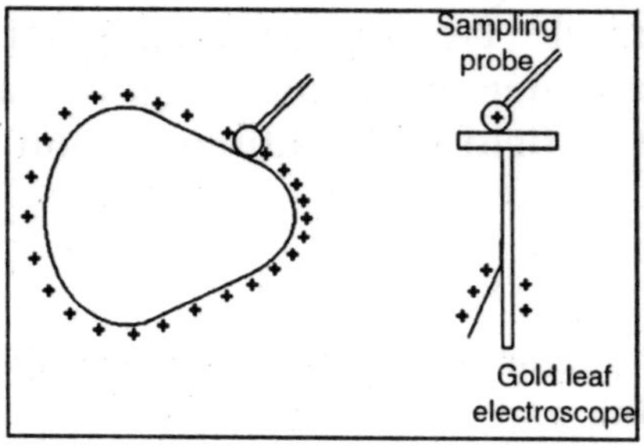

CAPACITANCE

For two conducting plates of area A separated by a distance d(m) with a voltage difference between them V the electric field between the plates is:

$$E = V/d$$

and the quantity of charge on each of the plates:

$$Q = A\,\sigma = \varepsilon_0\,AE = \varepsilon_0\,AV/d$$

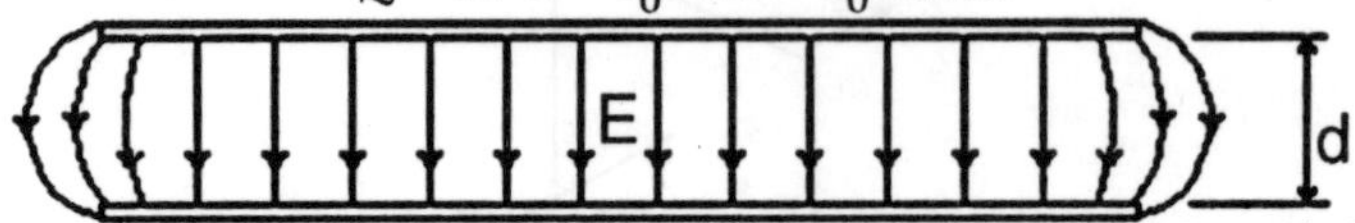

The proportionality between the quantity of charge Q and the voltage V is called the 'capacitance' C of the system. The unit of capacitance is the 'farad' with units of coulomb/volt. For the pair of parallel plates:

$$C = Q/V = \varepsilon_0\,A/d$$

This relation is not quite accurate because the electric field is not uniform near the edges of the plates and extends outwards from them a bit.

For an isolated sphere of radius a (m) the capacitance is $4\pi\varepsilon_0 a$– for a radius of 10mm this is about 1 picofarad. The capacitance for a number of simple geometric arrangements (for instance concentric spheres, concentric cylinders, parallel rods and rods near plane conductors) can be caculated analytically.

These can provide useful guidance for estimating capacitance in practical situations. Alternatively capacitance valus may be calculated by computer modelling calculations or, with suitable care, measured. The electrostatic energy U (joule) of a capacitor of C (farads) is:

$$U = 0.5\,CV^2$$

Poisson Equation:

$$\nabla^2 V = -\rho/\varepsilon_0$$

where ñ is the volume density of charge (C m^{-3}). If no charges are present then Poisson's equation becomes Laplace's equation:

$$\nabla^2 V = 0$$

There are a number of geometric forms for which the above equations can be solved analytically. Two and three dimensional finite element and finite difference computer modelling programs are available to find potential and electric field distributions within practical geometric arrangements.

DIELECTRICS

If a block of conductor were placed between the capacitor plates the field in the remaining gap would be increased. A similar effect occurs with insulating materials, but because charges cannot move through or over the surface of insulators the material becomes 'polarised'.

The atoms or molecules distort slightly under the electrical stress, polarise, and while these effects cancel out within the insulator they appear as surface charges on either side of the block of material. If an insulator is placed between the plates of a parallel plate capacitor the capacitance is increased.

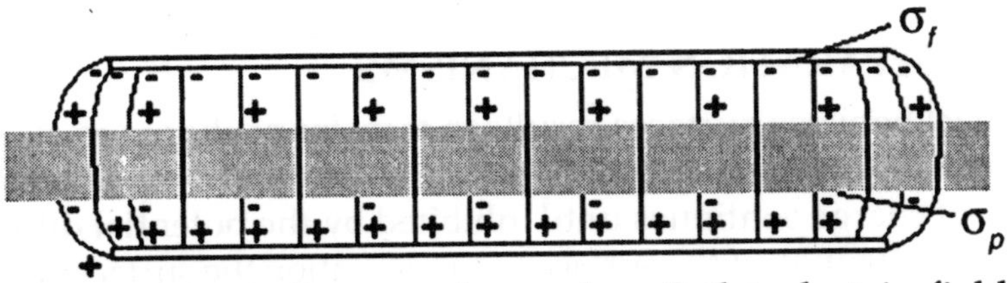

Applying Gauss's Law to the surface S the electric field in the insulator is:

$$E = (\sigma_f - \sigma_p)/\varepsilon_0$$

where σ_f and σ_p are the charge densities on the free surface of the capacitor plates and the surface of the insulator. The polarisation charges are proportional to the electric field and this is usually written:

$$P = X\,\varepsilon_0 E = \sigma_p$$

$$E = \sigma_f/\varepsilon_0(1 + X))$$

where X is the 'electric susceptibility' of the insulator/dielectric. This shows the reduction of the electric field within the insulator compared to that just outside.

The total charge on the capacitor plates is $\sigma_f A$ so the capacitance by $Q = CV$ becomes for a capacitor fully filled by an insulator:

$$C = \sigma_f A/V = AE\,(0(1 + X))/(E\,d)$$

$$C = \varepsilon_0 A\,(1 + X)/d$$

The increase in capacitance for this fully filled capacitor is the factor (1 + X) and this is usually called the 'Permittivity' of the material –*k*.

CHARGE MIGRATION AND DISSIPATION

Insulating materials have very few free electrical charges so only very small currents flow under the influence of an electric field. In more usual terms the volume 'resistivity' may be from 10^{12} to 10^{16} ohm m and the surface resistivity over 10^{10} ohms per square. Not only are there few charge carriers but their number and the velocity of their movement may depend in a non-linear way upon the electric field.

In this situation 'resistivity' is not a very useful parameter for characterising materials. Of more direct practical importance in the control of static is the timescale for static charge dissipation. If the timescale for charge dissipation is less than the timescale for charge generation then no significant charge levels or surface voltages will arise and there will be no static problems.

For uniform materials, such as liquids, it is expected that the time for charge dissipation $\tau(\sigma)$ can be related to the volume resistivity ρ (ohms) and permittivity k as:

$$\tau = k\, \varepsilon_0 \rho$$

CHARGE SEPARATION MECHANISMS

Contact between surfaces results in transfer of charge as electrons to match up energy levels of neighbouring atoms around the points of contact. The flow of charge continues until inhibited by the potential difference of the 'contact potential' of the surfaces. On separation the surfaces may retain the transferred charge.

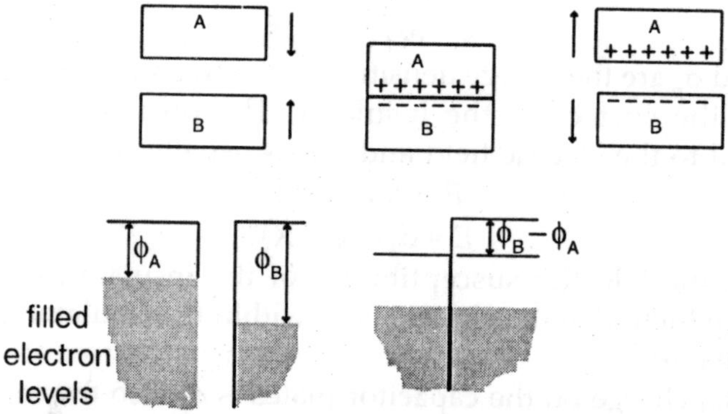

Charge transfer depends on differences in 'contact potential' and materials can be listed as a 'triboelectric series'. Charge separation depends on surface conditions. It is strongly affected by contamination and by the speed and pressure of rubbing actions.

Only in a few instances, for example in photocopiers and in some paint spray situation, are triboelectric effects used to control charging in a defined way. Charge separation also occurs when a liquid meets a conducting surface. Ions formed by dissociation in the fluid or from trace impurities will adsorb at the boundary surface.

This creates the so-called 'double layer' at the boundary. An appreciable charge can be adsorbed to the surface–which is matched by charge in the fluid. With relatively insulating fluids the charge in the volume can be swept away by fluid flow as a 'streaming current' which can be in the range 10^{-10} to 10^{-7}A. A double layer also occurs when water breaks away from a surface– and this is responsible for the formation of the highly charged mists that can arise when high pressure water jets hit surfaces.

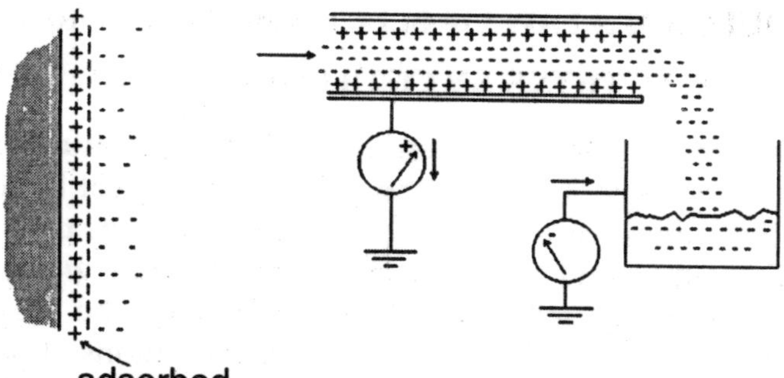

ELECTRICAL BREAKDOWN

If an electric field, in air at normal temperature and pressure, increases above about 3 MV m^{-1} then 'electrical breakdown' occurs. A loose electron or ion accelerates in the local electric field and if it picks up enough energy from the field before it hits another air molecule it will cause ionisation of that molecule.

This can lead to further ionisation and rapid growth of the number of ions present. This multiplication of ionisation can destroy the insulating properties of the air and lead to passage of a high current accompanied by the emission of heat and light and sound–this is a spark discharge.

If the electric field exceeds the critical value for ion multiplication in just a localised region, for example near a sharp point, while the average electric field over the gap to nearby conductors is quite low, then a localised discharge takes place with a fairly low current (up to perhaps a mA) between the electrodes.

This is a 'corona' discharge. If the electric field exceeds the breakdown strength of air over several mm, for example with an electrode of several mm radius of curvature, then the discharge involves more current and has a condensed luminous channel which does not reach to another electrode. This is a 'brush discharge'.

If the breakdown strength of air is exceeded near an electrode and also the average electric field over the gap to the other electrode exceeds about 500 kV m^{-1} then breakdown is likely to occur as a spark discharge between the electrodes.

The voltage for electrical breakdown depends on the geometry of the gap, on the gas and on the pressure. As the pressure or gap is reduced the breakdown voltage decreases to a minimum and then rises-following the Paschen Law curve. For air the minimum is about 300 volts. If a high electric field arises at the surface of a fluid the surface will distort and break up. This occurs at a lower electric field than that for air breakdown.

COULOMB'S LAW OF ELECTROSTATIC FORCES

CHARGE INTERACTIONS BY COULOMB'S LAW

Electrical interactions between charges govern much of physics, chemistry, and biology. They are the basis for chemical bonding, weak and strong. Salts dissolve in water to form the solutions of charged ions that cover three-quarters of the Earth's surface. Salt water forms the working fluid of living cells. pH and salts regulate the associations of proteins, DNA, cells, and colloids, and the conformations of biopolymers.

Nervous systems would not function without ion fluxes. Electrostatic interactions are also important in batteries, corrosion, and electroplating. Charge interactions obey Coulomb's law. When more than two charged particles interact, the energies are sums of coulombic interactions.

With these tools you can determine the electrostatic force exerted by one charged object on another, as when an ion interacts with a protein, DNA molecule, or membrane, or when a charged polymer changes conformation. Coulomb's law was discovered in careful experiments by H Cavendish, J Priestley, and CA Coulomb (1736–1806) on macroscopic objects such as magnets, glass rods, charged spheres, and silk cloths.

But Coulomb's law applies to a wide range of size scales, including atoms, molecules and biological cells. It states that the interaction energy *u(r)* between two charges in a vacuum is

$$u(r) = e\frac{q_1 q_2}{r},$$

where q_1 and q_2 are the magnitudes of the two charges, r is the distance separating them, and $e = 1/4\pi\varepsilon_0$ is a proportionality constant.

CHARGE INTERACTIONS BY LONG-RANGED

Uncharged atoms interact with each other through *short-ranged* interactions. That is, attractions between uncharged atoms are felt only when the two atoms are so close to each other that they nearly contact.

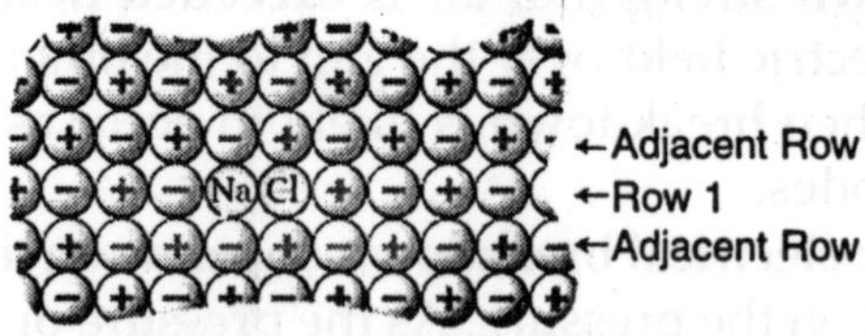

Fig. Crystalline Sodium Chloride Packs in a Cubic Array of Alternating Sodium and Chloride Ions

Short-ranged interactions typically diminish with distance as r^{-6} it suffices to note that charge interactions are very different: they are long-ranged. Coulombic interactions diminish with distance as r^{-1}. Counting lattice contacts is no longer sufficient. The difference between short-ranged and long-ranged interactions is much more profound than it may seem. The mathematics and physics of longranged and short-ranged interactions are very different.

A particle that interacts through short-ranged interactions feels only its nearest neighbors, so system energies can be computed by counting the nearest-neighbour contacts.

But when interactions involve charges, more distant neighbors contribute to the energies, so our methods of summing energies must become more sophisticated.

Electrostatic interactions are strong as well as long-ranged. The electrostatic repulsion is about 1036 times stronger than the gravitational attraction between two protons. Strong, short-ranged nuclear forces are required to hold the protons together against their electrostatic repulsions in atomic nuclei.

When charge builds up on a macroscopic scale, violent events may result, such as lightning and thunder storms, and explosions in oil tankers. Because charge interactions are so strong, matter does not sustain charge separations on macroscopic scales: macroscopic matter is neutral.

The transfer of less than a thousand charges is enough to cause the static electricity that you observe when you rub your shoes on a carpet. In solution, too, imbalances are usually less than thousands of charges, but because this is one part in 1021, it means that you can assume that solutions have charge neutrality—no net charge of one sign—to a very high degree of approximation. Although macroscopic charge separations aren't common in ordinary solutions, charge separations on microscopic scales.

CHARGE INTERACTIONS BY WEAKER IN MEDIA

Charges interact more weakly when they are in liquids than when they are in a vacuum. To varying degrees, liquids can be *polarized*. This means that if two fixed charges, say a negative charge A and a positive charge B, are separated by a distance r in a liquid, then the intervening molecules of the liquid tend to reorient their charges.

The positive charges in the medium orient toward A, and the negative charges toward B, to shield and weaken the interaction between A and B. This weakening of the coulombic interactions between A and B is described by the *dielectric constant* of the medium. Media that can be readily polarized shield charge interactions strongly: they have high dielectric constants.

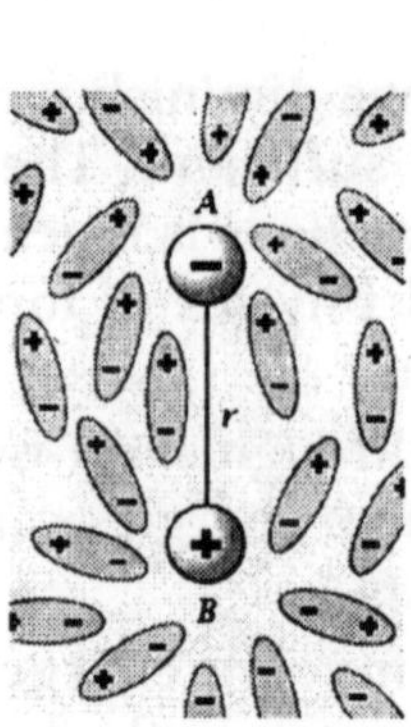

Fig. Fixed Charges *A* and *B* Cause Dipoles in the Surrounding Liquid to Orient Partly, Which Causes some *Shielding*, a Reduction of the Interaction Between *A* and *B*.

The medium as a *polarizable isotropic continuum.* In treating a medium as a continuum, we neglect its atomic structure and focus on its larger-scale properties. In treating a medium as *isotropic,* we assume that its polarizability is the same in all directions. By treating a medium as *polarizable,* we assume that the charge redistributes in response to an electric field, even if the medium is neutral overall.

The more polarizable the medium, the greater is the reduction of an electrostatic field across it. The reduction factor is the *dielectric constant* D, a scalar dimensionless quantity. When charges q_1 and q_2 are separated by a distance r in a medium of dielectric constant D, their coulombic interaction energy is

$$u(r) = \frac{eq_1q_2}{Dr}.$$

Coulombic interactions are traditionally described as an energy $u(r)$, but when dielectric constants are involved, $u(r)$ is a free energy, because the dielectric constant is temperature dependent. The polarizability of a medium arises from several factors.

First, if the medium is composed of molecules that have a *permanent dipole moment,* a positive charge at one end and a negative charge at the other, then applying an electric field tends to orient the dipoles of the medium in a direction that opposes the field.

Second, the polarizability of a medium can arise from the polarizabilities of the atoms or molecules composing it, even if they have no permanent dipole moment. Atoms or molecules that lack permanent dipoles have electronic *polarizability,* a tendency of nuclear or electronic charge distributions to shift slightly within the atom, in response to an electric field. The electronic polarizabilities of hydrocarbons and other nonpolar substances are the main contributors to their dielectric constants ($D \approx 2$).

Third, polarizabilities can also arise from networks of hydrogen bonds in liquids such as water. If there are many alternative hydrogen-bond donors and acceptors, a shift in the pattern of hydrogen bonding is not energetically costly, so the networks of hydrogen bonds in hydrogen-bonded liquids are labile when electric fields are applied.

THE BJERRUM LENGTH

To show the effect of the polarizability of the medium, let's first compute a useful quantity called the *Bjerrum length –B*, defined as the charge separation at which the coulomb energy between ions *u(r)* just equals the thermal energy *RT*.

Substitute ℓ_B for r and RT for u in $u(r) = eq_1q_2/Dr$ and solve for ℓ_B forsingle charges $q_1 = q_2 = e$:

$$\ell_B = \frac{ee^2N}{DRT}.$$

For a temperature $T = 298.15$, in a vacuum or air ($D = 1$),

$$\ell_B = \frac{1.386 \times 10^{-4}\,\mathrm{J\,m\,mol^{-1}}}{(8.314\,\mathrm{J\,K^{-1}mol^{-1}} \times 298.15K)} = 560 Å.$$

560 Å is a relatively long distance, many times the diameter of an atom. When two charges are much closer together than this distance, they feel a strong repulsive or attractive interaction, much larger than the thermal energy *RT*. But when two charges are much further apart than this distance, their interactions are weaker than *RT*, and they are more strongly directed by Brownian motion than by their coulombic interactions.

Dividing $u(r) = eq_1q_2/Dr$ by $(RT = eq_1q_2/D\ell_B)$ gives

$$\frac{u}{RT} = \frac{\ell_B}{r}.$$

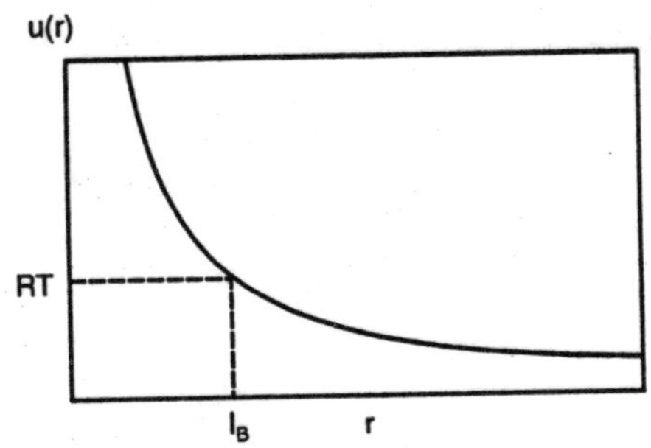

Fig. The Bjerrum Length *B* is Defined as the Distance *r* at Which two Unit Charges Interact with an Energy *u(r)* Equal to *RT*.

So far, our calculations for ionic interactions have involved simple arrangements of charges. When several charges are involved, or when charges have complex spatial arrangements, more sophisticated methods are needed to treat them. We now develop some useful tools that lead to Gauss's law and Poisson's equation.

ELECTROSTATIC FORCES ADD LIKE VECTORS

Vectors and Coulomb's Law

Coulomb's law may be expressed either in terms of energy $u(r) = eq_1q_2/Dr$, or force $f = -(\partial u/\partial r) = eq_1q_2/Dr^2$.

The difference is that energies are scalar quantities, which add simply, while forces are vectors that add in component form. Either way, an important fundamental law is the superposition principle: electrostatic energies and forces are both additive. The total electrostatic force on a particle is the vector sum of the electrostatic forces from all other particles.

$$f = \frac{eq_Aq_B}{Dr^2}.\frac{r}{r},$$

Where r/r indicates a vector pointing in the direction of vector r, along the line connecting the particles, but with unit length. f is the force acting in the same direction.

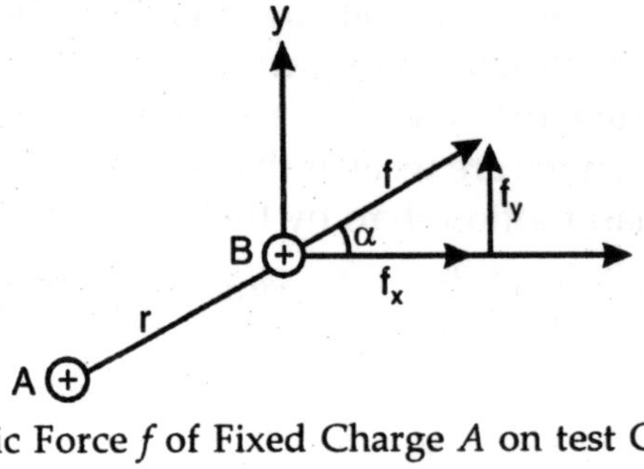

Fig. The Electrostatic Force f of Fixed Charge A on test Charge B is a Vector with Components fx and fy Along the x- and y-axes, Respectively.

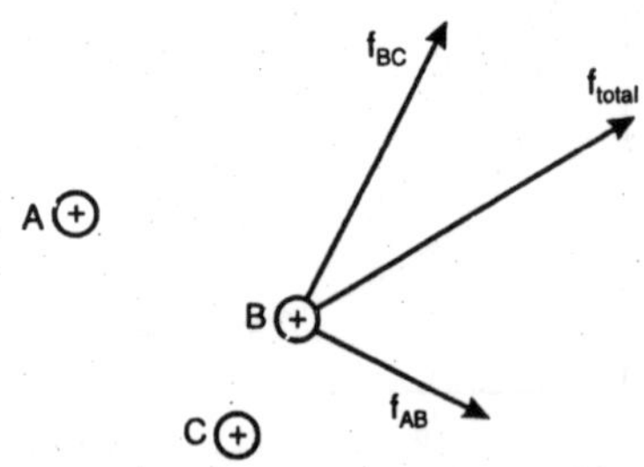

Fig. The Force on the Test Charge B is the Vector Sum from the two Fixed Charges A and C.

The component of the force f in the x-direction is described by using the component of the vector r between the charges, $rx = r \cos \alpha$, where α is the angle between f and the x-axis. The x-direction component of the force is

$$f_x = \left(\frac{q_A q_B}{4\pi\varepsilon_0 D r^2} \right) \cos \alpha.$$

The y-direction component is:

$$f_y = \left(\frac{q_A q_B}{4\pi\varepsilon_0 D r^2} \right) \sin \alpha.$$

ELECTRIC FIELD

In physics, the space surrounding an electric charge or in the presence of a time-varying magnetic field has a property called an electric field (that can also be equated to electric flux density). This electric field exerts a force on other electrically charged objects. The concept of electric field was introduced by Michael Faraday.

The electric field is a vector field with SI units of newtons per coulomb ($N\ C^{-1}$) or, equivalently, volts per meter ($V\ m^{-1}$). The strength of the field at a given point is defined as the force that would be exerted on a positive test charge of +1 coulomb placed at that point; the direction of the field is given by the direction of that force. Electric fields contain electrical energy with energy density proportional to the square of the field intensity. The electric field is to charge as gravitational acceleration is to mass and force density is to volume.

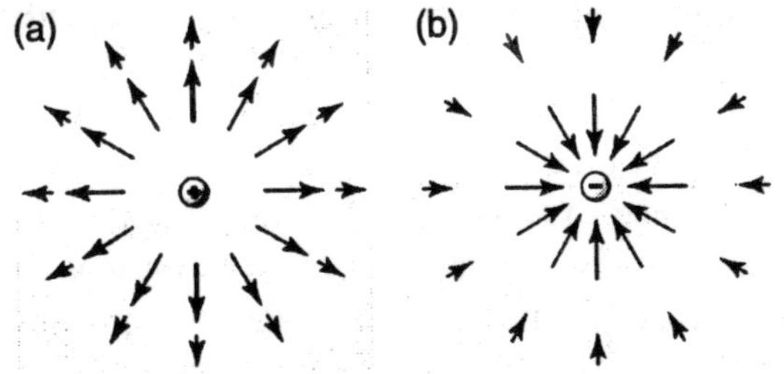

Fig. The Electric Field is indicated by Vectors Pointing.

(a) Away from a Positive Charge

(b) Toward a Negative Charge.

The Field Represents the Force Per Unit Charge that acts on a Positive Test Charge Put at Each Particular Position in Space. A moving charge has not just an electric field but also a magnetic field, and in general the electric and magnetic fields are not completely separate phenomena; what one observer perceives as an electric field, another observer in a different frame of reference perceives as a mixture of electric and magnetic fields.

For this reason, one speaks of "electromagnetism" or "electromagnetic fields." In quantum mechanics, disturbances in the electromagnetic fields are called photons, and the energy of photons is quantized. A stationary charged particle in an electric field experiences a force proportional to its charge given by the equation,

$$F = q\left[-\nabla\phi - \frac{\partial A}{\partial t}\right]$$

where the magnetic flux density is given by,

$$B = \nabla \times A$$

and where $-\nabla\phi$ is the Coulomb force.

Electric charge is a characteristic of some subatomic particles, and is quantized when expressed as a multiple of the so-called elementary charge e. Electrons by convention have a charge of –1, while protons have the opposite charge of +1. Quarks have a fractional charge of –1/3 or +2/3. The antiparticle equivalents of these have the opposite charge. There are other charged particles.

In general, same-sign charged particles repel one another, while different-sign charged particles attract. This is expressed quantitatively in Coulomb's law, which states the magnitude of the repelling force is proportional to the product of the two charges, and weakens proportionately to the square of the distance.

The electric charge of a macroscopic object is the sum of the electric charges of its constituent particles. Often, the net electric charge is zero, since naturally the number of electrons in every atom is equal to the number of the protons, so their charges cancel out. Situations in which the net charge is non-zero are often referred to as static electricity.

Even when the net charge is zero, it can be distributed non-uniformly (e.g., due to an external electric field), and then the material is said to be polarized, and the charge related to the polarization is known as bound charge (while the excess charge brought from outside is called free charge).

An ordered motion of charged particles in a particular direction (in metals, these are the electrons) is known as electric current. The discrete nature of electric charge was proposed by Michael Faraday in his electrolysis experiments, then directly demonstrated by Robert Millikan in his oil-drop experiment.

The SI unit for quantity of electricity or electric charge is the coulomb, which represents approximately 1.60×10^{19} elementary charges (the charge on a single electron or proton). The coulomb is defined as the quantity of charge that has passed through the cross-section of an electrical conductor carrying one ampere within one second. The symbol *Q* is often used to denote a quantity of electricity or charge.

The quantity of electric charge can be directly measured with an electrometer, or indirectly measured with a ballistic galvanometer. Formally, a measure of charge should be a multiple of the elementary charge e (charge is quantized), but since it is an average, macroscopic quantity, many orders of magnitude larger than a single elementary charge, it can effectively take on any real value.

In some contexts it is meaningful to speak of fractions of a charge; e.g. in the charging of a capacitor. If the charged particle can be considered a point charge, the electric field is defined as the force it experiences per unit charge:

$$E = \frac{F}{q}$$

where as:

F is the electric force experienced by the particle;

q is its charge;

E is the electric field wherein the particle is located;

Taken literally, this equation only defines the electric field at the places where there are stationary charges present to experience it. The force exerted by another charge q will alter the source distribution, which means the electric field in the presence of q differs from itself in the absence of q. However, the electric field of a given source distribution remains defined in the absence of any charges with which to interact.

This is achieved by measuring the force exerted on successively smaller *test charges* placed in the vicinity of the source distribution. By this process, the electric field created by a given source distribution is defined as the limit as the test charge approaches zero of the force per unit charge exerted there upon.

$$E = \lim_{q \to 0} \frac{F}{q}$$

This allows the electric field to be dependent on the source distribution alone. As is clear from the definition, the direction of the electric field is the same as the direction of the force it would exert on a positively-charged particle, and opposite the direction of the force on a negatively-charged particle. Since like charges repel and opposites attract the electric field tends to point away from positive charges and towards negative charges

Electric Fields have Fluxes

The electric field flux has the important general property that it is independent of the size and shape of any balloon that fully surrounds all the charges, as we will show. This generality leads to some important methods for simplifying the problem of computing complex electrostatic fields. Positive and negative charges are the sources and sinks of electrical fields.

Consider some constellation of positive and negative charges fixed at certain positions in space. Just as we did for fluids we invent an imaginary balloon that we place around the charges. We want to determine the flux of the electric field through the imaginary balloon surface.

The electric field flux Φ is defined as the integral of D times the electric field E over all the surface area elements ds of the balloon, where D is the dielectric constant. Because E is proportional to $1/D$ in a uniform medium the flux Φ is independent of D.

$$\Phi = \int_{surface} DE.ds,$$

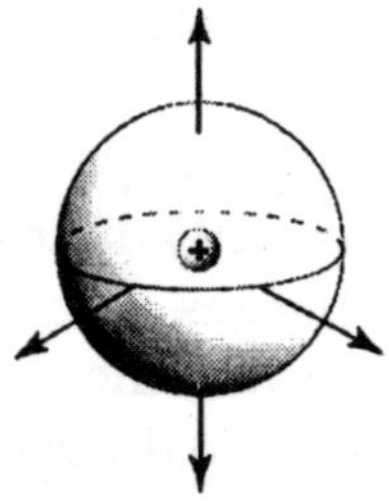

Fig. The Vector Electric Field E is Normal to a Sphere Surrounding a Point Charge located at the Centre of the Sphere

Equation is general and applies to any constellation of fixed charges, and to any shape and position of an imaginary balloon around them. Many problems of interest have a special simplicity or symmetry that allows you to simplify Equation.

For example, suppose that the charge constellation is just a single point charge q and that the imaginary balloon is just a sphere of radius r centered on the point charge. We are interested in computing the electric field flux from the point charge through the imaginary spherical surface.

The electrostatic field of a point charge is spherically symmetrical—the magnitude of the field E is the same through the north pole as it is through the south pole or through the equator, or through any other patch of surface of a given area. So you can drop the vector notation and express the field as $E(r)$, which depends only on r. Because E doesn't differ from one surface patch to another, the integral in Equation becomes a simple product.

That is, E is parallel to ds everywhere, so $E \cdot ds = Eds$ and the flux integral becomes:

$$\Phi = DE \int ds = DE(r) \times \text{(total area of sphere)}.$$

Use $E(r) = q/(4\pi\varepsilon_0 Dr^2)$ from Equation and $4\pi r^2$ for the surface area of a sphere, to get

$$\begin{aligned}\Phi &= DE(r)(4\pi r^2)\\ &= \left(\frac{q}{4\pi\varepsilon_0 r^2}\right)(4\pi r^2)\\ &= \frac{q}{\varepsilon_0},\end{aligned}$$

where q is the charge at the centre of the sphere.

Two profound implications of Equation lead to a general method for calculating electrostatic forces and fields. First, the right-hand side of Equation does not depend on r. For the electric flux through a spherical container, the r^{-2} dependence of E cancels with the r^2 dependence of the spherical surface area. Because of this cancelation, the electric field flux out of a sphere is independent of its radius. It doesn't matter how large you choose your imaginary balloon to be, the flux is the same through spherical balloons of any radius.

POISSON'S EQUATION

In mathematics, Poisson's equation is a partial differential equation with broad utility in electrostatics, mechanical engineering and theoretical physics. It is named after the French mathematician, geometer and physicist Siméon-Denis Poisson. The Poisson equation is

$$\Delta\varphi = f$$

where Δ is the Laplace operator, and f and ö are real or complex-valued functions on a manifold. When the manifold is Euclidean space, the Laplace operator is often denoted as ∇^2 and so Poisson's equation is frequently written as:

$$\nabla^2\varphi = f$$

In three-dimensional Cartesian coordinates, it takes the form:

$$\left(\frac{\partial^2}{\partial x^2}+\frac{\partial^2}{\partial y^2}+\frac{\partial^2}{\partial z^2}\right)\varphi(x,y,z) = f(x,y,z).$$

For vanishing f, this equation becomes Laplace's equation:

$$\Delta\varphi = 0$$

The Poisson equation may be solved using a Green's function; a general exposition of the Green's function for the Poisson equation is given in the article on the screened Poisson equation. There are various methods for numerical solution. The relaxation method, an iterative algorithm, is one example.

ELECTROSTATICS

One of the principal cornerstones of electrostatics is the posing and solving of problems that are described by the Poisson equation. Finding φ for some given f is an important practical problem, since this is the usual way to find the electric potential for a given charge distribution.

The derivation of Poisson's equation in electrostatics follows. SI units are used and Euclidean space is assumed. Starting with Gauss' law for electricity in a differential control volume, we have:

$$\nabla . D = \rho$$

∇ means to take the divergence.

D is the electric displacement field.

ρ is the charge density.

Assuming the medium is linear, isotropic, and homogeneous, then:

$$D = \varepsilon E = \varepsilon_r \varepsilon_0 E$$

ε is the permittivity of the medium.

E is the electric field.

ε_0 is the vacuum permittivity.

ε_r is the relative permittivity of the medium.

By substitution and division, we have:

$$\nabla . E = \frac{\rho}{\varepsilon_r \varepsilon_0}$$

In the absence of a changing magnetic field, B, Faraday's law of induction gives:

$$\nabla \times E = -\frac{\partial B}{\partial t} = 0$$

$\times$ is the cross product.

t is time.

Since the curl of the electric field is zero, it is defined by a scalar electric potential field, φ

$$E = -\nabla \varphi$$

Eliminating E by substitution, we have a form of the Poisson equation:

$$\nabla . \nabla \varphi = \nabla^2 \varphi = -\frac{\rho}{\varepsilon_r \varepsilon_0}$$

Solving Poisson's equation for the potential requires knowing the charge density distribution. If the charge density is zero, then Laplace's equation results. If the charge density follows a Boltzmann distribution, then the Poisson-Boltzmann equation results.

The Poisson-Boltzmann equation plays a role in the development of the Debye-Hückel theory of dilute electrolyte solutions. The magnetic field not varying in time, the same Poisson equation arises even if it does vary in time, as long as the Coulomb gauge is used.

However, in this more general context, computing φ is no longer sufficient to calculate E, since the latter also depends on the magnetic vector potential, which must be independently computed.

POTENTIAL OF A GAUSSIAN CHARGE DENSITY

If there is a spherically symmetric Gaussian charge density $\rho(r)$:

$$\rho(r) = \frac{Q}{\sigma^3 \sqrt{2\pi}^3} e^{-r^2/(2\sigma^2)},$$

where Q is the total charge, then the solution $\varphi(r)$ of Poisson's equation with $\varepsilon_r = 1$,

$$\nabla^2 \varphi = -\frac{\rho}{\varepsilon_0},$$

is given by

$$\varphi(r) = \frac{1}{4\pi\varepsilon_0} \frac{Q}{r} erf\left(\frac{r}{\sqrt{2}\sigma}\right).$$

where erf(x) is the error function. This solution can be checked explicitly by a careful manual evaluation of $\nabla^2\varphi$. Note that, for r much greater than σ, erf(x) approaches unity and the potential $\phi(r)$ approaches the point charge potential,

$$\frac{1}{4\pi\varepsilon_0} \frac{Q}{r}$$

as one would expect. The erf function approaches 1 extremely fast as its argument increase; in practice for $r > 3\sigma$ the relative error is smaller than 1/1000.

LAPLACE'S EQUATION

Laplace's equation is a partial differential equation named after Pierre-Simon Laplace who first studied its properties. The solutions of Laplace's equation are important in many fields of science, notably the fields of electromagnetism, astronomy, and fluid dynamics, because they describe the behaviour of electric, gravitational, and fluid potentials.

The general theory of solutions to Laplace's equation is known as potential theory. In the study of heat conduction, the Laplace equation is the steady-state heat equation. In three dimensions, the problem is to find twice-differentiable real-valued functions, φ of real variables, *x*, *y*, and *z*, such that:

$$\frac{\partial^2 \varphi}{\partial x^2} + \frac{\partial^2 \varphi}{\partial y^2} + \frac{\partial^2 \varphi}{\partial z^2} = 0.$$

This is often written as:

$$\nabla^2 \varphi = 0$$

or

$$\text{div grad } \varphi = 0,$$

where div is the divergence, and grad is the gradient, or

$$\Delta \varphi = 0,$$

where Δ is the Laplace operator.

Solutions of Laplace's equation are called harmonic functions.

If the right-hand side is specified as a given function, $f(x, y, z)$, i.e.

$$\Delta \varphi = f$$

then the equation is called "Poisson's equation." Laplace's equation and Poisson's equation are the simplest examples of elliptic partial differential equations. The partial differential operator, ∇^2, or Δ, (which may be defined in any number of dimensions) is called the Laplace operator, or just the Laplacian.

BOUNDARY CONDITIONS

The Dirichlet problem for Laplace's equation consists of finding a solution φ on some domain *D* such that φ on the boundary of *D* is equal to some given function. Since the Laplace operator appears in the heat equation, one physical interpretation of this problem is as follows: fix the temperature on the boundary of the domain and wait until the temperature in the interior doesn't change anymore; the temperature distribution in the interior will then be given by the solution to the corresponding Dirichlet problem.

The Neumann boundary conditions for Laplace's equation specify not the function φ itself on the boundary of *D*, but its normal derivative. Physically, this corresponds to the construction of a potential for a vector field whose effect is known at the boundary of *D* alone.

Solutions of Laplace's equation are called harmonic functions; they are all analytic within the domain where the equation is satisfied. If any two functions are solutions to Laplace's equation (or any linear homogenous differential equation), their sum (or any linear combination) is also a solution. This property, called the principle of superposition, is very useful, e.g., solutions to complex problems can be constructed by summing simple solutions.

LAPLACE EQUATION IN TWO DIMENSIONS

The Laplace equation in two independent variables has the form

$$\varphi_{xx} + \varphi_{yy} = 0.$$

Analytic Functions

The real and imaginary parts of a complex analytic function both satisfy the Laplace equation. That is, if $z = x + iy$, and if

$$f(z) = u(x, y) + iv(x, y),$$

then the necessary condition that $f(z)$ be analytic is that the Cauchy-Riemann equations be satisfied:

$$u_x = v_y,\ v_x = -u_y.$$

It follows that:

$$u_{yy} = (-v_x)_y = -(v_y)_x = -(u_x)_x.$$

Therefore u satisfies the Laplace equation. A similar calculation shows that v also satisfies the Laplace equation.

Conversely, given a harmonic function, it is the real part of an analytic function, $f(z)$ (at least locally). If a trial form is

$$f(x) = \varphi(x, y) + i\psi(x, y),$$

then the Cauchy-Riemann equations will be satisfied if we set

$$\psi_x = \psi_y,\ \psi_y = \varphi x.$$

This relation does not determine ø, but only its increments:

$$d\psi = -\varphi y dx + \varphi_x dy.$$

The Laplace equation for φ implies that the integrability condition for ψ is satisfied:

$$\psi_{xy} = \psi_{yx}.$$

and thus ψ may be defined by a line integral. The integrability condition and Stokes' theorem implies that the value of the line integral connecting two points is independent of the path. The resulting pair of solutions of the Laplace equation are called conjugate harmonic functions. This construction is only valid locally, or provided that the path does not loop around a singularity. For example, if r and θ are polar coordinates and

$$\varphi = \log r,$$

then a corresponding analytic function is

$$f(x) = \log z = \log r + i\theta.$$

However, the angle θ is single-valued only in a region that does not enclose the origin. The close connection between the Laplace equation and analytic functions implies that any solution of the Laplace equation has derivatives of all orders, and can be expanded in a power series, at least inside a circle that does not enclose a singularity. This is in sharp contrast to solutions of the wave equation, which generally have less regularity.

There is an intimate connection between power series and Fourier series. If we expand a function f in a power series inside a circle of radius R, this means that

$$f(z)=\sum_{n=0}^{\infty}c_n z^n,$$

with suitably defined coefficients whose real and imaginary parts are given by

$$c_n = a_n + ib_n.$$

Therefore:

$$f(z)=\sum_{n=0}^{\infty}[a_n r^n \cos n\theta - b_n r^n \sin n\theta]$$

$$+\ i\sum_{n=1}^{\infty}[a_n r^n \sin n\theta + b_n r^n \cos n\theta],$$

which is a Fourier series for f.

Fluid Flow

Let the quantities u and v be the horizontal and vertical components of the velocity field of a steady incompressible, irrotational flow in two dimensions. The condition that the flow be incompressible is that

$$u_x + v_y = 0,$$

and the condition that the flow be irrotational is that

$$\nabla \times V = v_x - u_y = 0.$$

If we define the differential of a function ψ by

$$d\psi = v\,dx - u\,dy,$$

then the incompressibility condition is the integrability condition for this differential: the resulting function is called the stream function because it is constant along flow lines.

The first derivatives of ψ are given by:

$$\psi x = v,\ \psi y = -u,$$

and the irrotationality condition implies that ψ satisfies the Laplace equation. The harmonic function ψ that is conjugate to ψ is called the velocity potential.

The Cauchy-Riemann equations imply that:

$$\varphi_x = -u,\ \varphi_y = -v.$$

Thus every analytic function corresponds to a steady incompressible, irrotational fluid flow in the plane. The real part is the velocity potential, and the imaginary part is the stream function.

According to Maxwell's equations, an electric field (u,v) in two space dimensions that is independent of time satisfies

$$\nabla \times (u,v) = v_x - u_y = 0,$$

and

$$\nabla.(u,v) = \rho$$

where ρ is the charge density. The first Maxwell equation is the integrability condition for the differential

$$d\varphi = -u\,dx - v\,dy,$$

so the electric potential φ may be constructed to satisfy

$$\varphi_x = -u,\ \varphi_y = -v.$$

The second of Maxwell's equations then implies that

$$\varphi_{xx} + \varphi_{yy} = -\rho,$$

which is the Poisson equation.

It is important to note that the Laplace equation can be used in three-dimensional problems in electrostatics and fluid flow just as in two dimensions.

LAPLACE EQUATION IN THREE DIMENSIONS

Fundamental Solution

A fundamental solution of Laplace's equation satisfies

$$\Delta u = u_{xx} + u_{yy} + u_{zz} = -\delta(x - x', y - y', z - z'),$$

where the Dirac delta function δ denotes a unit source concentrated at the point (x', y', z'). No function has this property, but it can be thought of as a limit of functions whose integrals over space are unity, and whose support (the region where the function is non-zero) shrinks to a point. The definition of the fundamental solution thus implies that, if the Laplacian of u is integrated over any volume that encloses the source point, then

$$\iiint_V div \nabla u dV = -1.$$

the Laplace equation is unchanged under a rotation of coordinates, and hence we can expect that a fundamental solution may be obtained among solutions that only depend upon the distance r from the source point. If we choose the volume to be a ball of radius a around the source point, then Gauss' divergence theorem implies that;

$$-1 = \iiint_V div \nabla u dV = \iint_S u_r dS = 4\pi a^2 u_r(a).$$

It follows that

$$u_r(r) = -\frac{1}{4\pi r^2},$$

on a sphere of radius r that is centered around the source point, and hence

$$u = \frac{1}{4\pi r}.$$

a similar argument shows that in two dimensions

$$u = \frac{-\log r}{4\pi}.$$

Green's Function

A Green's function is a fundamental solution that also satisfies a suitable condition on the boundary S of a volume V. For instance, $G(x, y, z; x', y', z')$ may satisfy

$$\nabla.\nabla G = -\delta(x - x', y - y', z - z')$$

$$G = 0 \text{ if } (x, y, z) \text{ on } S.$$

Now if u is any solution of the Poisson equation in V:

$$\nabla.\nabla\, u = -f,$$

and u assumes the boundary values g on S, then we may apply Green's identity, (a consequence of the divergence theorem) which states that

$$\iiint_V [G\nabla.\nabla u - u\nabla.\nabla G] dV = \iiint_V \nabla.[G\nabla u - u\nabla G] dV$$

$$= \iiint_S [Gu_n - uG_n] dS.$$

The notations u_n and G_n denote normal derivatives on S. In view of the conditions satisfied by u and G, this result simplifies to;

$$u(x', y', z') = \iiint_V G f dV + \iint_S G_n g dS.$$

Thus the Green's function describes the influence at of the data f and g. For the case of the interior of a sphere of radius a, the Green's function may be obtained by means of a reflection the source point P at distance ñ from the centre of the sphere is reflected along its radial line to a point P' that is at a distance

$$\rho' = \frac{a^2}{\rho}.$$

Note that if P is inside the sphere, then P' will be outside the sphere. The Green's function is then given by

$$\frac{1}{4\pi R}-\frac{a}{4\pi\rho R'},$$

where R denotes the distance to the source point P and R' denotes the distance to the reflected point P'. A consequence of this expression for the Green's function is the Poisson integral formula. Let ρ, θ, and φ be spherical coordinates for the source point P. Here θ denotes the angle with the vertical axis, which is contrary to the usual American mathematical notation, but agrees with standard European and physical practice. Then the solution of the Laplace equation inside the sphere is given by

$$u(P)=\frac{1}{4\pi}a^3\left(1-\frac{\rho^2}{a^2}\right)\iint\frac{g(\theta',\varphi')\sin\varphi' d\theta' d\varphi'}{(a^2+\rho^2-2a\rho\cos\Theta)^{3/2}},$$

where

$$\cos\Theta=\cos\varphi\cos\varphi'+\sin\varphi\sin\varphi'\cos(\theta-\theta').$$

A simple consequence of this formula is that if u is a harmonic function, then the value of u at the centre of the sphere is the mean value of its values on the sphere. This mean value property immediately implies that a non-constant harmonic function cannot assume its maximum value at an interior point.

Electrostatic Potential

Electrostatics is the part of physics that describes interactions between stationary charges. You are probably familiar with Coulomb's Law, the central law of electrostatics. This law says that two charged particles exert a force on each other equal to:

$$F = q_1 q_2 / r_{12}^2$$

The electrostatic force, F, is proportional to the product of the charges on the two particles, q_1 and q_2, and inversely proportional to the *square* of the distance separating the particles, r_{12}. Another important characteristic of a charged system is its potential energy, PE. Potential energy is created by electrostatic interactions between charge particles and is equal to:

$$PE = q_1 q_2 / r_{12}$$

Notice that this formula looks nearly the same as Coulomb's Law. The only difference is that potential energy is inversely proportional to the distance between charges, while the Coulomb force is inversely proportional to the *square* of the distance.

The most useful quantity for our purposes is the *electrostatic potential*. This quantity is related to PE as follows: the electrostatic potential created by a system of charges at a particular point in space, (x, y, z), is equal to the change in potential energy that occurs when a +1 ion is introduced at this point.

This definition can be made clearer with the aid of the following pictures. Imagine a molecule consisting of an electron density cloud and several positively charged nuclei.

Now suppose we want to know the electrostatic potential this molecule creates at point (x, y, z). We can obtain the potential by introducing a +1 charge at (x, y, z) and calculating the change in energy. The new picture looks like this:

The change in energy is simply the potential energy created by interactions between the +1 charge and the charges in the molecule. We can calculate this energy by calculating

$$(+1)(q_{\text{molecule}})/r$$

for each charge in the molecule, q_{molecule}, and adding up all of these energies. This energy is the molecule's electrostatic potential.

Electrostatic potential is both a molecular property and a spatial property. It depends on what charges exist in the molecule and how they there are distributed. It also depends on what point (x, y, z) we choose to investigate.

If we select a point where the +1 charge is attracted by the molecule, the potential will be negative at this point. On the other hand, if we select a point where the +1 charge is repelled, the potential will be positive.

Molecules contain many charged particles, nuclei and electrons, and the net impact of these particles on the +1 "probe" can only be determined by a computer. However, since we know that potential energy and distance are inversely related, it is likely that the molecular charge(s) closest to the +1 particle have the largest effect.

For example, the following diagram shows an ionic compound consisting of three ions. It is likely that the potential in the immediate vicinity of each ion is determined largely by this ion, and the more distant ions have relatively small effect.

On the other hand, the potential in any region that is near two or more ions must be determined by a careful calculation.This kind of behaviour is seen in practically every system. Positive particles, like atomic nuclei or polyatomic cations, are surrounded by regions of positive potential.

Likewise, negative particles, like polyatomic anions, are surrounded by regions of negative potential. In a few pages I will show you how to use electrostatic potentials to make qualitative statements about atomic charges. When I do this, I will assume that the potential in a given region is controlled by the "local" atom, so a positive potential will indicate a positively charged "local" atom and a negative potential will indicate a negatively charged "local" atom.

Electrostatic Potential Maps

The same Spartan 02 computer programme that is used to construct molecular models, and calculate model electron density clouds, is also be used to calculate model electrostatic potentials. The process is complicated because the programme must calculate the interaction between the +1 probe charge and every part of the electron density cloud, and it is necessary to divide the cloud into many small pieces to do this accurately.

The display of electrostatic potential data is another difficult problem. We cannot look at the potential everywhere simultaneously, so we must select some points of interest first, and display the potential at these points.

Spartan selects the points located on the molecule's size density surface (0.002 au isodensity surface) and calculates the potential at these points. Then Spartan assigns colors to these points according to their potential. The resulting rainbow-colored surface is called a *potential map.*

Default Colour Scale

Spartan uses colour to show the value of the potential at different points on the map. Each potential corresponds to a unique colour, and you "read" a map by noticing where different colors appear.

Spartan can select colors automatically or it can select colors according to instructions that you give it. When Spartan02 is allowed to select colors automatically, it is uses a *default colour scale* to equate different potentials with different colors.

First, the programme surveys all of the potentials on the map, and finds the most negative and most positive potentials. These regions are colored red and dark blue, respectively. Intermediate potentials are then assigned colors according to the standard colour spectrum:

red (most negative) < yellow < green < light blue
< dark b'ue (most positive)

The potentials on this map range from –46 to +48 kcal/mol. The red region near oxygen shows the location of the most negative potential, –46 kcal/mol (this region attracts the +1 probe). The dark blue regions near the hydrogens show the locations of the most positive potential, +48 kcal/mol (these regions repel the +1 probe).

The green region shows the location of the mean potential, the average of the two extremes. On this map, the mean potential is +1 kcal/mol. The yellow region and light blue regions show the locations of potentials halfway between the mean (green) and extremes (red and dark blue). On this map, yellow corresponds to –22 kcal/mol and light blue corresponds to +24 kcal/mol.

Standard Colour Scale

When we use the default colour scale, we know that the red and dark blue regions correspond to the *extreme* values of the electrostatic potential, but how can we know what these extreme potentials are? Different maps have different extremes. As a result, if we don't know the numerical value of the extreme potentials, we might be fooled into thinking that we are looking at large extremes (and large differences between extremes) when we are not.

Consider the potential map of propane, $CH_3CH_2CH_3$, shown below. This map is colored according to the default scale. It turns blue near every hydrogen, and yellow-orange around every carbon (the "cavity" at the top of the map is bordered by all three carbons and is red). These colors seem to suggest that the potential swings wildly between hydrogen (medium-dark blue) and carbon (yellow-red).

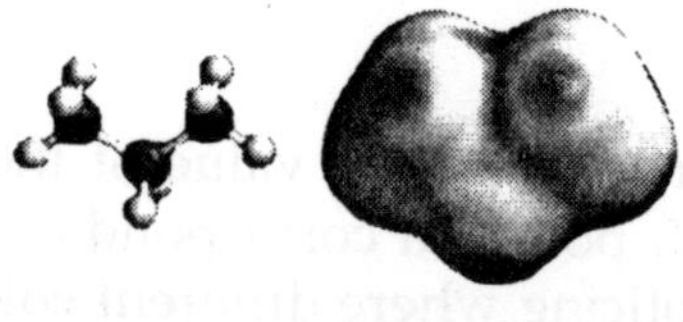

Fig. Potential Map of Propane, $CH_3CH_2CH_3$

Now examine the potential map of 1-propanol, $CH_3CH_2CH_2OH$, shown below. This map is also colored according to its default colour scale. Extreme potentials are located near the OH group. The $CH_3CH_2CH_2$ group is almost uniformly green. This suggests that the potential does not change much between hydrogen and carbon in this molecule.

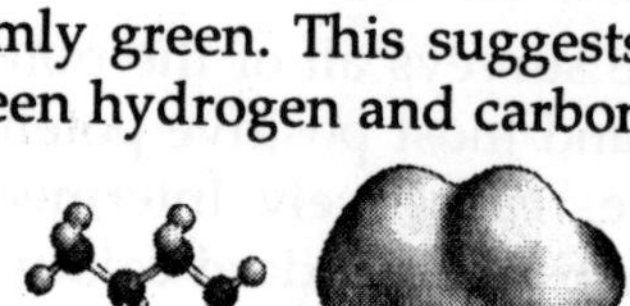

Fig. Potential Map of 1-propanol, $CH_3CH_2CH_2OH$

There does not appear to be any consistency in these maps. The $CH_3CH_2CH_2$ group appears to generate extreme potentials in propane, but a nearly constant potential in 1-propanol. If we carry this idea to its logical conclusion, we would say that the $CH_3CH_2CH_2$ groups are radically different in these molecules. Unfortunately, this conclusion is contradicted by several other types of data (bond distances, isodensity surfaces, electronegativity data, and so on), so we must seek another explanation. The answer lies in the numerical value of the extreme potentials. The potentials on the propane map swing from –22 to +4 kcal/mol, while the potentials on the 1-propanol map cover a much wider range, –48 to +44 kcal/mol. Since both maps use the same number of colors, each colour on 1–

propanol's map covers a wider range of potentials. This is why the entire $CH_3CH_2CH_2$ group has roughly the same colour on the 1–propanol map.

We can avoid these problems by giving the values of the extreme potentials for every map. When the extreme potentials are known, the meaning of other colors can be inferred as follows:

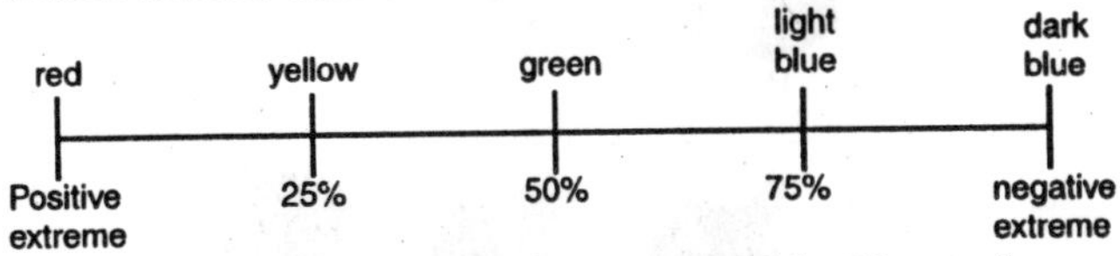

Green always corresponds to a potential halfway between the two extremes (red/dark blue). Yellow and light blue split the difference between the mean (green) and the extremes (red/dark blue).

Applying this to the 1-propanol map (red = –48 and blue = +44 kcal/mol) gives green = –2, yellow = –25, and light blue = +21 kcal/mol. Notice that the entire potential range of propane's map, –22 to +4 kcal/mol, would be assigned two colors, green and yellow, on 1-propanol's map.

Another solution that we can also use is to colour the maps of ***neutral*** molecules using a *standard colour scale.* My definition of this scale places the extremes at –40 (red) and +40 kcal/mol (dark blue). This means that intermediate colors correspond to modest potentials of –20 (yellow), 0 (green), and +20 (light blue) kcal/mol.

The potential maps of propane and 1-propanol colored according to my standard colour scale (remember that we have not changed the potentials, just the colors assigned to each potential). Now the $CH_3CH_2CH_2$ group looks roughly the same in both molecules.

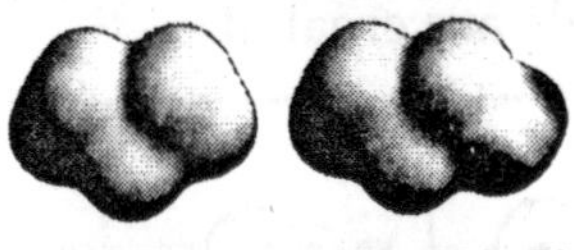

Fig. Potential Maps of Propane, $CH_3CH_2CH_3$, and 1-propanol, $CH_3CH_2CH_2OH$ (Standard Colour Scale)

The standard colour scale is useful, but it is not entirely free of problems. Suppose the potentials on a given map lie *outside* the standard extremes, that is, suppose some potentials are less than –40 or greater than +40 kcal/mol.

POLAR H–X BONDS

The isodensity surfaces can be used to get information about H-X bond polarity. This information can be obtained much more easily from a potential map.

The maps are colored using the standard colour scale, and the molecules are oriented so that one H is always on the left side of the map.

The potential *near hydrogen* is very negative in LiH, close to neutral in BH_3 and CH_4, and increasingly positive in NH_3, H_2O, and HF. As we expect the potentials closely track the charge on hydrogen.

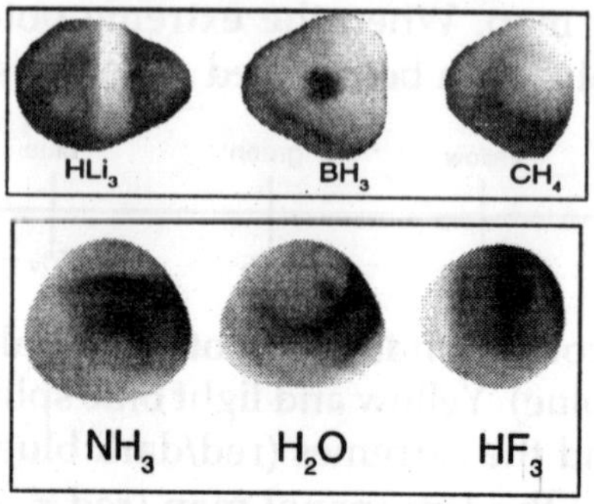

Fig. Potential Maps of HLi_3, BH_3, CH_4, NH_3, H_2O, HF_3.

Although the potential near hydrogen varies systemat-ically with the electronegativity of X, the potential near X does not. N and O are surrounded by more negative potentials than F, an unexpected result since F is more electronegative.

This strange behaviour is due to a combination of factors, one of which is the non-spherical shape of X. It is not hard to see how a non-spherical electron density cloud might create unusual electrostatic potentials.

Consider the cartoon atoms shown below. If the nucleus is at the centre of a spherical cloud, the potential is zero everywhere outside the atom.

(A). However, if the nucleus is set to one side of the cloud

(B). Or the cloud is non-spherical

(C). Unusual potentials are created.

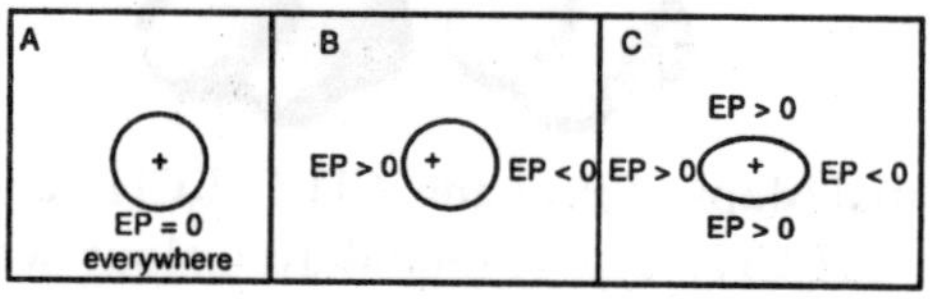

Chemists refer to the regions of positive and negative potential as *electron-poor* and *electron-rich*, respectively. Thus, the red regions in the NH_3 and H_2O maps appear to be electron-rich.

POLAR C–X BONDS

Carbon is only slightly more electronegative than hydrogen. Therefore, we expect C–F, C–O, anc' C–N bonds to be polar, just like the corresponding H–F, H–O, and H–N bonds. Unfortunately, polar bonds involving carbon are almost impossible to detect using potential maps.

Methyl propionate, $CH_3CH_2C(=O)OCH_3$, furnishes us with a good example of this phenomenon. Two carbons in this molecule are bonded to oxygen. The *methyl* carbon makes one CO bond, O–CH_3, and the *carbonyl* carbon makes three CO bonds, C(=O)–O.

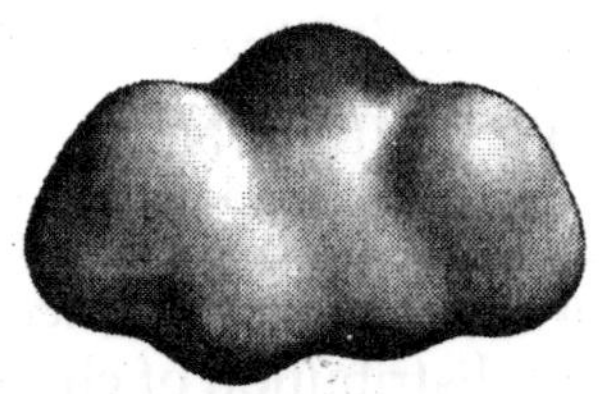

Fig. Potential Map of Methyl Propionate, $CH_3CH_2C(=O)OCH_3$

Since all of the CO bonds are polar, all of them transfer electron density from carbon to oxygen and we expect to find large positive potentials around these carbons. The following potential map of methyl propionate standard colour scale) shows that this does not happen. The potential near each carbon is very small.

Two factors account for these observations. First, the "local" atom principle breaks down near carbon whenever it is sandwiched between several other atoms. Notice that the carbonyl carbon in methyl propionate lies in between two oxygens, both of which create large local negative potentials. These negative potentials offset the positive potentials created by carbon.

A second important factor is atom size. Larger atoms have larger size density surfaces. This means we calculate and display potential at regions farther away from the atom. Since potential falls off as $1/r$, a potential map will show smaller potentials near a large atom than it will near a small atom, other factors being equal.

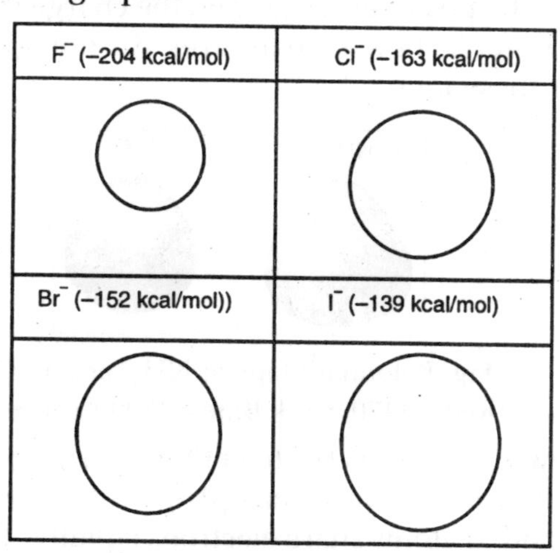

Fig. Potential Maps of Halide Anions (Red = –200, Dark blue = –120 Kcal/mol): F^-, Cl^-, Br^-, I^-.

All of these ions carry identical charges of –1, but the potential is most negative around the smallest anion, fluoride. As ion and map size increase, the map's potential becomes steadily less negative.

CHARGE DISTRIBUTION IN IONS

Electrostatic potential maps can be used to analyse charge distributions in ions. Although an ion's Lewis structure necessarily contains a (formally) charged atom, the actual distribution of charge may be quite different.

To begin, consider the potential maps of CH_3^- and HO^- shown below (colour scale = –200 to –120 kcal/mol). The Lewis structures of these ions assign a –1 formal charge to C and O, respectively. The potential maps support these charge assignments in that the most negative potentials appear near the charged atoms. Lewis structures appear to give accurate charge information for these ions.

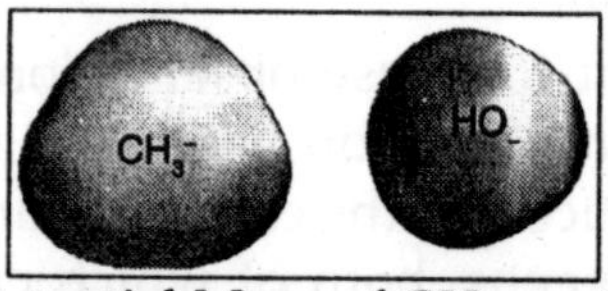

Fig. Potential Maps of CH_3^- and HO^-
(Red = –200, Dark Blue = –120 Kcal/mol)

Next, consider the potential maps of NH_4^+ and H_3O^+ shown below (colour scale = +120 to +200 kcal/mol). The Lewis structures of these ions assign a +1 formal charge to N and O, respectively, but the potential maps do not support these charge assignments. According to the maps, the most positive potentials appear near the hydrogens. We must conclude that the hydrogens carry the positive charge, and the charge is spread equally over all of the hydrogens. Lewis structures do not give accurate charge information for these ions.

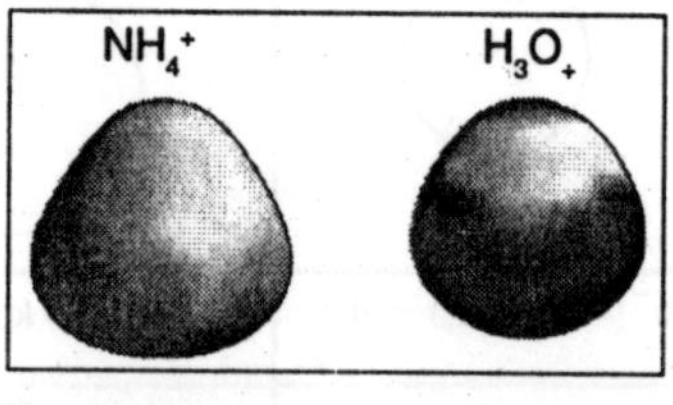

Fig. Potential Maps of NH_4^+ and H_3O^+
(Red = +120, Dark Blue = +200 Kcal/mol)

Why do Lewis structures fail to get the charges right in these cations? The answer to this in the assumption underlying formal charge assignments: bonded atoms share electrons equally. The O in H_3O^+ carries *a* + 1 charge if it shares bonding electrons equally with the neighboring H. We know, however, that O is more electronegative than H (and O^+ is more electronegative still), so this assumption cannot be correct. Formal charges can give the wrong picture.

So far we have focused on charge location, the red region on an anion's map and the blue region on a cation's map. What about the other colors/ regions on these maps? Does the red region near oxygen in H_3O^+ indicate that oxygen is negative?

Before you take the bait and say, "yes, a red O must be negatively charged", consider this: a cation repels a +1 charge no matter how the +1 charge approaches the cation. In other words, all of the regions on the potential map of H_3O^+ are positive *regardless of colour*. The red colour on the map corresponds to a positive potential, as do all of the other colors.

We cannot guess the charges of atoms near any of these other colors. Potential maps are especially valuable tools for identifying and characterizing ionic resonance hybrids.

Partial bonding in an ionic resonance hybrid can spread the ion's charge over two or more atoms. This type of charge is known as a *delocalized charge*. Each charged atom carries only a portion of the delocalized charge, and these atoms create noticeably smaller "local" potentials than atoms with full charges.

One way to detect delocalized charge is to compare the potential map of a suspected ionic resonance hybrid with the map of an ion whose charge is concentrated on a single atom of the same type (a charge that is concentrated on one atom is known as a *localized charge*).

For example, we can tell if the negative charge in $CH_3CH_2CH_2C(=O)O^-$ is delocalized by comparing the potential map of this compound with the potential map of a localized ion like $CH_3CH_2CH_2CH_2O^-$. To make this comparison meaningful, we use the same colour scale for both maps, and we compare charged atoms (oxygens) of similar size.

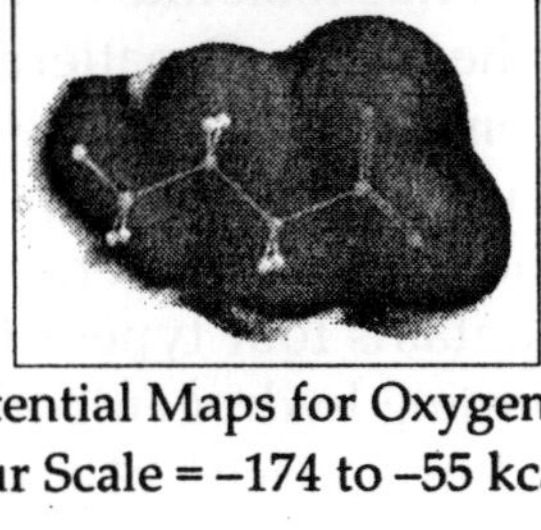

Fig. Potential Maps for Oxygen Anions
(Colour Scale = –174 to –55 kcal/mol)

The two oxygen atoms in the suspect compound produce significantly smaller negative potentials (yellow-orange) than the oxygen atom in the reference ion (red).

In addition, both oxygens in the suspect compound produce similar potentials. These observations suggest that the charge in $CH_3CH_2CH_2$ $C(=O)O^-$ is delocalized over both oxygens:

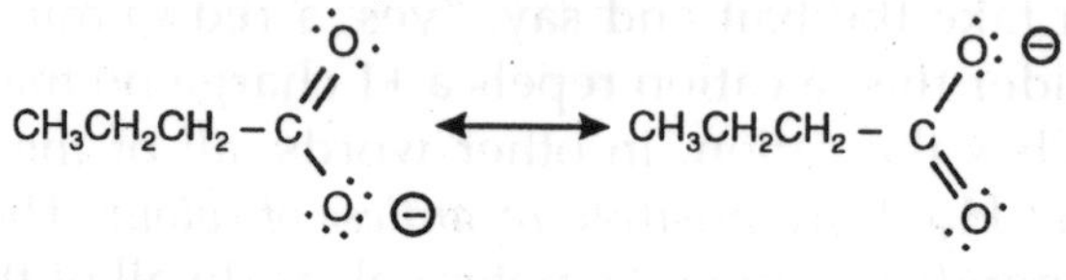

HYDROGEN BONDS

One of the most important nonbonded interactions that can be studied with potential maps is the *hydrogen bond* or *H bond*. This kind of "bond" occurs when two atoms, a positively charged hydrogen atom and a negatively charged non-hydrogen atom, interact electrostatically.

Hydrogen bonds occur most often between neutral molecules. The interacting atoms in these molecules carry fractional charges, and this type of hydrogen bond can be considered a type of dipole-dipole interaction.

The fractionally charged atoms responsible for most hydrogen bonds normally generate significant "local" potentials. Experience shows that hydrogen bonds typically involve a hydrogen atom whose "local" potential is $\geq +40$ kcal/mol, and a non-hydrogen partner whose "local" potential is ≤ -40 kcal/mol.

Hydrogens bonded to either oxygen or fluorine typically generate "local" potentials of +40 kcal/mol or more. These atoms always form strong hydrogen bonds with a suitable partner. Hydrogens bonded to certain types of nitrogen can also generate large "local" potentials, and many hydrogen-bonded systems are based on this type of hydrogen. Hydrogens bonded to carbon rarely generate large potentials, and CH_n groups rarely, if ever, form strong hydrogen bonds.

On the other side of the ledger, only nitrogen and oxygen atoms seem to generate large negative "local" potentials in neutral compounds. Therefore, most hydrogen bonds look like O–H... X or N–H... X, where X = N or O. An example of the O–H... O pattern can be found in a cup of water. The two oxygen atoms belong to different water molecules, and we can draw the hydrogen bonded water molecules as HO–H... OH_{-2}.

Several examples of the N–H... X pattern appear in double-stranded DNA. Each DNA strand contains four types of rings (called nucleotides) that are responsible for forming hydrogen bonds with rings in the other strand. The hydrogen bonding capacity of each nucleotide also controls

DNA synthesis, and insures that a newly synthesized strand contains the proper sequence of nucleotides.

Two nucleotides, cytosine and guanine, can form three hydrogen bonds simultaneously. The other nucleotides, adenine and thymine, can form two. All five hydrogen bonds follow the N–H... X pattern, and potential maps show that the potentials near these hydrogens are all greater than +40 kcal/mol. Potential maps of cytosine and guanine are shown below.

The molecules have been pushed apart so that their respective potential maps can be compared. The maps show that all three of the N–H groups responsible for hydrogen bonds generate large "local" positive potentials. The maps also show that the nitrogen and oxygen partners generate large "local" negative potentials.

Each molecule generates a complementary pattern of positive and negative potentials so that, when they approach, three stabilizing interactions will be created.

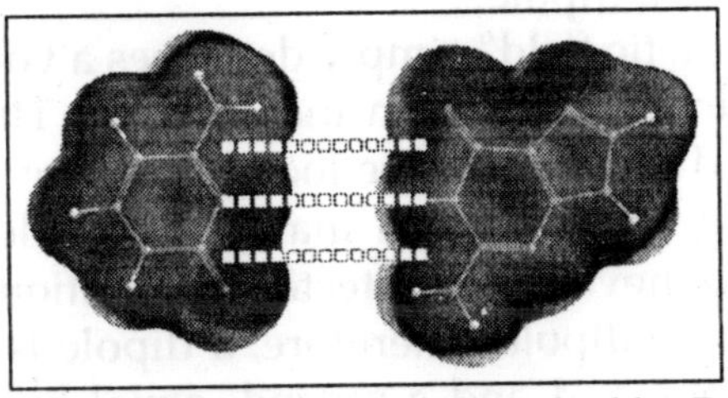

Fig. Hydrogen Bond Partners are Identified by Dashed Lines

Chapter 8

Magnetic Force

Magnets are very common items in the workplace and household. Uses of magnets range from holding pictures on the refrigerator to causing torque in electric motors. Most people are familiar with the general properties of magnets but are less familiar with the source of magnetism. The traditional concept of magnetism centers around the magnetic field and what is know as a dipole.

The term "magnetic field" simply describes a volume of space where there is a change in energy within that volume. This change in energy can be detected and measured. The location where a magnetic field can be detected exiting or entering a material is called a magnetic pole. Magnetic poles have never been detected in isolation but always occur in pairs, hence the name dipole. Therefore, a dipole is an object that has a magnetic pole on one end and a second, equal but opposite, magnetic pole on the other.

A bar magnet can be considered a dipole with a north pole at one end and south pole at the other. A magnetic field can be measured leaving the dipole at the north pole and returning the magnet at the south pole. If a magnet is cut in two, two magnets or dipoles are created out of one. This sectioning and creation of dipoles can continue to the atomic level. Therefore, the source of magnetism lies in the basic building block of all matter the atom.

Magnetism is a force generated in matter by the motion of electrons within its atoms. Magnetism and electricity represent different aspects of the force of electromagnetism, which is one part of Nature's fundamental electroweak force. The region in space that is penetrated by the imaginary lines of magnetic force describes a magnetic field. The strength of the magnetic field is determined by the number of lines of force per unit area of space.

Magnetic fields are created on a large scale either by the passage of an electric current through magnetic metals or by magnetized materials called magnets. The elemental metals-iron, cobalt, nickel, and their solid solutions or alloys with related metallic elements-are typical materials that respond strongly to magnetic fields.

Unlike the all-pervasive fundamental force field of gravity, the magnetic force field within a magnetized body, such as a bar magnet, is polarized-that is, the field is strongest and of opposite signs at the two extremities or poles of the magnet.

THE SOURCE OF MAGNETISM

All matter is composed of atoms, and atoms are composed of protons, neutrons and electrons. The protons and neutrons are located in the atom's nucleus and the electrons are in constant motion around the nucleus. Electrons carry a negative electrical charge and produce a magnetic field as they move through space. A magnetic field is produced whenever an electrical charge is in motion. The strength of this field is called the magnetic moment.

This may be hard to visualize on a subatomic scale but consider electric current flowing through a conductor. When the electrons (electric current) are flowing through the conductor, a magnetic field forms around the conductor. The magnetic field can be detected using a compass.

The magnetic field will place a force on the compass needle, which is another example of a dipole.Since all matter is comprised of atoms, all materials are affected in some way by a magnetic field. However, not all materials react the same way.

HISTORY OF MAGNETISM

The history of magnetism dates back to earlier than 600 b.c., but it is only in the twentieth century that scientists have begun to understand it, and develop technologies based on this understanding. Magnetism was most probably first observed in a form of the mineral magnetite called lodestone, which consists of iron oxide-a chemical compound of iron and oxygen. The ancient Greeks were the first known to have used this mineral, which they called a magnet because of its ability to attract other pieces of the same material and iron.

The Englishman William Gilbert was the first to investigate the phenomenon of magnetism systematically using scientific methods. He also discovered that the Earth is itself a weak magnet. Early theoretical investigations into the nature of the Earth's magnetism were carried out by the German Carl Friedrich Gauss.

Quantitative studies of magnetic phenomena initiated in the eighteenth century by Frenchman Charles Coulomb (1736-1806), who established the inverse square law of force, which states that the attractive force between two magnetized objects is directly proportional to the product of their individual fields and inversely proportional to the square of the distance between them.

Danish physicist Hans Christian Oersted first suggested a link between electricity and magnetism. Experiments involving the effects of magnetic and electric fields on one another were then conducted by Frenchman Andre Marie Ampere and Englishman Michael Faraday, but it was the Scotsman, James Clerk Maxwell, who provided the theoretical foundation to the physics of electromagnetism in the nineteenth century by showing that electricity and magnetism represent different aspects of the same fundamental force field.

Then, in the late 1960s American Steven Weinberg and Pakistani Abdus Salam, performed yet another act of theoretical synthesis of the fundamental forces by showing that electromagnetism is one part of the electroweak force.

The modern understanding of magnetic phenomena in condensed matter originates from the work of two Frenchmen: Pierre Curie, the husband and scientific collaborator of Madame Marie Curie, and Pierre Weiss. Curie examined the effect of temperature on magnetic materials and observed that magnetism disappeared suddenly above a certain critical temperature in materials like iron. Weiss proposed a theory of magnetism based on an internal molecular field proportional to the average magnetization that spontaneously align the electronic micromagnets in magnetic matter.

The present day understanding of magnetism based on the theory of the motion and interactions of electrons in atoms (called quantum electrodynamics) stems from the work and theoretical models of two Germans, Ernest Ising and Werner Heisenberg (1901-1976). Werner Heisenberg was also one of the founding fathers of modern quantum mechanics.

Magnetism arises from two types of motions of electrons in atoms-one is the motion of the electrons in an orbit around the nucleus, similar to the motion of the planets in our solar system around the sun, and the other is the spin of the electrons around its axis, analogous to the rotation of the Earth about its own axis.

The orbital and the spin motion independently impart a magnetic moment on each electron causing each of them to behave as a tiny magnet. The magnetic moment of a magnet is defined by the rotational force experienced by it in a magnetic field of unit strength acting perpendicular to its magnetic axis. In a large fraction of the elements, the magnetic moment of the electrons cancel out because of the Pauli exclusion principle, which states that each electronic orbit can be occupied by only two electrons of opposite spin.

However, a number of so-called transition metal atoms, such as iron, cobalt, and nickel, have magnetic moments that are not cancelled; these elements are, therefore, common examples of magnetic materials. In these transition metal elements the magnetic moment arises only from the spin of the electrons.

In the rare earth elements (that begin with lanthanum in the sixth row of the Periodic Table of Elements), however, the effect of the orbital motion of the electrons is not cancelled, and hence both spin and orbital motion contribute to the magnetic moment.

Examples of some magnetic rare earth elements are: cerium, neodymium, samarium, and europium. In addition to metals and alloys of transition and rare earth elements, magnetic moments are also observed in a wide variety of chemical compounds involving these elements. Among the common magnetic compounds are the metal oxides, which are chemically bonded compositions of metals with oxygen.

The Earth's geomagnetic field is the result of electric currents produced by the slow convective motion of its liquid core in accordance with a basic law of electromagnetism which states that a magnetic field is generated by the passage of an electric current. According to this model, the Earth's core should be electrically conductive enough to allow generation and transport of an electric current.

The geomagnetic field generated will be dipolar in character, similar to the magnetic field in a conventional magnet, with lines of magnetic force lying in approximate planes passing through the geomagnetic axis. The principle of the compass needle used by the ancient mariners involves the alignment of a magnetized needle along the Earth's magnetic axis with the imaginary south pole of the needle pointing towards the magnetic north pole of the Earth. The magnetic north pole of the Earth is inclined at an angle of 11 degrees away from its geographical north pole.

TYPES OF MAGNETISM

Five basic types of magnetism have been observed and classified on the basis of the magnetic behaviour of materials in response to magnetic fields at different temperatures. These types of magnetism are: ferromagnetism, ferrimagnetism, antiferromagnetism, paramagnetism, and diamagnetism.

Ferromagnetism and ferrimagnetism occur when the magnetic moments in a magnetic material line up spontaneously at a temperature below the so-called Curie temperature, to produce net magnetization. The magnetic moments are aligned at random at temperatures above the Curie point, but become ordered, typically in a vertical or, in special cases, in a spiral (helical) array, below this temperature. In a ferromagnet magnetic moments of equal magnitude arrange themselves in parallel to each other. In a ferrimagnet, on the other hand, the moments are unequal in magnitude and order in an antiparallel arrangement.

When the moments are equal in magnitude and ordering occurs at a temperature called the Neel temperature in an antiparallel array to give no net magnetization, the phenomenon is referred to as antiferromagnetism. These transitions from disorder to order represent classic examples of phase transitions. Another example of a phase transition is the freezing of the disordered molecules of water at a critical temperature of 32°F (0°C) to form the ordered structure of ice.

The magnetic moments-referred to as spins-are localized on the tiny electronic magnets within the atoms of the solid. Mathematically, the electronic spins are equal to the angular momentum (the rotational velocity times the moment of inertia) of the rotating electrons.

The spins in a ferromagnetic or a ferrimagnetic single crystal undergo spontaneous alignment to form a macroscopic (large scale) magnetized object. Most magnetic solids, however, are not single crystals, but consist of single crystal domains separated by domain walls. The spins align within a domain below the Curie temperature, independently of any external magnetic field, but the domains have to be aligned in a magnetic field in order to produce a macroscopic magnetized object.

This process is effected by the rotation of the direction of the spins in the domain wall under the influence of the magnetic field, resulting in a displacement of the wall and the eventual creation of a single large domain with the same spin orientation.

Paramagnetism is a weak form of magnetism observed in substances which display a positive response to an applied magnetic field. This response is described by its magnetic susceptibility per unit volume, which is a dimensionless quantity defined by the ratio of the magnetic moment to the magnetic field intensity.

Paramagnetism is observed, for example, in atoms and molecules with an odd number of electrons, since here the net magnetic moment cannot be zero. Diamagnetism is associated with materials that have a negative magnetic susceptibility. It occurs in nonmagnetic substances like graphite, copper, silver and gold, and in the superconducting state of certain elemental and compound metals.

The negative magnetic susceptibility in these materials is the result of a current induced in the electron orbits of the atoms by the applied magnetic field. The electron current then induces a magnetic moment of opposite sign to that of the applied field. The net result of these interactions is that the material is shielded from penetration by the applied magnetic field.

MEASUREMENT OF MAGNETIC FIELD

The magnetic field or flux density is measured in metric units of a gauss (G) and the corresponding international system unit of a tesla (T). The magnetic field strength is measured in metric units of oersteds (Oe) and international units of amperes per meter (A/m). Instruments called gaussmeters and magnetometers are used to measure the magnitude of magnetic fields.

One form of the gaussmeter that is used commonly in the laboratory consists of a current carrying semiconducting element called the Hall probe, which is placed perpendicular to the magnetic field being measured. As a consequence of the so-called Hall effect, a voltage perpendicular to the field and to the current is generated in the probe. This induced voltage is proportional to the magnetic field being measured and can be simply measured using a voltmeter.

Magnetometers are extremely sensitive magnetic field detectors. In one commonly used form the magnetic force is detected by means of a sensitive electronic balance. In this instrument the magnetic substance is placed on one arm of a balance, which in turn is placed in a magnetic field. The magnetic force on the sample is then determined by the weight required to balance the force generated by the magnetic field.

The most sensitive magnetometer in a modern physics laboratory utilizes a magnetic sensing element called the SQUID (which stands for Superconducting QUantum Interference Device). A SQUID consists of an extremely thin electrically resistive junction (called a Josephson junction) between two superconductors.

Superconductors are materials which undergo a transition at low temperatures to a state of zero electrical resistance and nearly complete exclusion of magnetic fields.

In its direct current mode of operation, a SQUID is first cooled down to its superconducting state, and then a current is passed through it while the voltage across the junction is monitored. When the junction senses a magnetic field, the flow of current is altered due to an interference phenomenon at the quantum level between two electron wave fronts through the junction, resulting in a change in voltage.

Interference is a phenomenon that occurs generally due to the mixing of two wave fronts; the waves add up in some regions and cancel out in others depending on the location of the crest and trough of each wave in space. For example, the interference between the sound waves from two simultaneously played musical instruments tuned at somewhat different frequencies results in the occurrence of beats or modulations in the sound intensity.

A variation of the SQUID magnetometer is the SQUID gradiometer which measures differences in magnetic fields at different positions. Using this type of instrument magnetic field variations in the femtotesla (10^{-15} tesla) range can be detected. Devices of this type have been used to map the tiny magnetic signals from the human brain.

APPLICATIONS OF MAGNETISM

Electromagnets are utilized as key components of transformers in power supplies that convert electrical energy from a wall outlet into direct current energy for a wide range of electronic devices, and in motors and generators. High field superconducting magnets (where superconducting coils generate the magnetic field) provide the magnetic field in MRI (magnetic resonance imaging) devices that are now used extensively in hospitals and medical centers.

Magnetic materials that are difficult to demagnetize are used to construct permanent magnets. Permanent magnet applications are in loudspeakers, earphones, electric meters, and small motors. A loudspeaker consists of a wire carrying an alternating current. When the wire is in the magnetic field of the permanent magnet it experiences a force that generates a sound wave by alternate compression and rarefaction of the surrounding air when the alternating frequency of the current is in the audible range.

The more esoteric applications of magnetism are in the area of magnetic recording and storage devices in computers, and in audio and video systems. Magnetic storage devices work on the principle of two stable magnetic states represented by the 0 and 1 in the binary number system. Floppy disks have dozens of tracks on which data can be digitally written in or stored by means of a write-head and then accessed or read by means of a read-head.

A write-head provides a strong local magnetic field to the region through which the storage track of the disk is passed. The read-head senses stray magnetic flux from the storage track of the disk as it passes over the head. Another example of digital magnetic storage and reading is the magnetic strip on the back of plastic debit and credit cards. The magnetic strip contains identification data which can be accessed through, for example, an automatic teller machine.

SOME CURRENT RESEARCH TRENDS IN MAGNETISM

Ideally pure magnetic systems have provided the most extensively investigated models of the large scale collective behaviour of atoms and electrons that occur in the vicinity of the critical point of phase transitions. Fascinating effects caused by the intentional introduction of impurities and defects into random locations in the atomic lattice of a magnetic material.

For example, these random magnetic systems display transitions to states of order that have no counterparts in pure systems, because pure systems are, by necessity, always close to thermodynamic equilibrium or stability. For these reasons there is now intense interest and research activity in disordered systems, and random magnets provide ideal model systems for such investigations.

An area of intense current activity centers around the search for a likely magnetic pairing force in the high temperature ceramic superconductors that were discovered in 1987 by the German-Swiss team of Georg Bednorz and Karl Alexander Muller. A superconductor achieves a zero resistance state by means of a force field that pairs up the conducting electrons within its atoms.

The new ceramic materials are antiferromagnets in their undoped state, but on doping start to superconduct at temperatures that are over 182°F (83°C) warmer than conventional pure metal and alloy superconductors.

The effects of extremely high magnetic fields on the properties of condensed matter continues to be an area of high interest. New research areas, such as the search and study of magnetism in organic matter, and the study of diamagnetism and novel magnetic effects in the recently synthesized nanometer-sized (a nanometer is equal to 10^{-9} meter) carbon tubes, are of increasing interest to physicists and material scientists.

DIAMAGNETIC, PARAMAGNETIC, AND FERROMAGNETIC MATERIALS

When a material is placed within a magnetic field, the magnetic forces of the material's electrons will be affected. This effect is known as Faraday's Law of Magnetic Induction. However, materials can react quite differently to the presence of an external magnetic field.

This reaction is dependent on a number of factors, such as the atomic and molecular structure of the material, and the net magnetic field associated with the atoms. The magnetic moments associated with atoms have three origins. These are the electron orbital motion, the change in orbital motion caused by an external magnetic field, and the spin of the electrons.

In most atoms, electrons occur in pairs. Electrons in a pair spin in opposite directions. So, when electrons are paired together, their opposite spins cause their magnetic fields to cancel each other. Therefore, no net magnetic field exists. Alternately, materials with some unpaired electrons will have a net magnetic field and will react more to an external field. Most materials can be classified as diamagnetic, paramagnetic or ferromagnetic.

Diamagnetic metals have a very weak and negative susceptibility to magnetic fields. Diamagnetic materials are slightly repelled by a magnetic field and the material does not retain the magnetic properties when the external field is removed.

Diamagnetic materials are solids with all paired electron resulting in no permanent net magnetic moment per atom. Diamagnetic properties arise from the realignment of the electron orbits under the influence of an external magnetic field. Most elements in the periodic table, including copper, silver, and gold, are diamagnetic.

Paramagnetic metals have a small and positive susceptibility to magnetic fields. These materials are slightly attracted by a magnetic field and the material does not retain the magnetic properties when the external field is removed. Paramagnetic properties are due to the presence of some unpaired electrons, and from the realignment of the electron orbits caused by the external magnetic field. Paramagnetic materials include magnesium, molybdenum, lithium, and tantalum.

Ferromagnetic materials have a large and positive susceptibility to an external magnetic field. They exhibit a strong attraction to magnetic fields and are able to retain their magnetic properties after the external field has been removed. Ferromagnetic materials have some unpaired electrons so their atoms have a net magnetic moment. They get their strong magnetic properties due to the presence of magnetic domains. In these domains, large numbers of atom's moments (10^{12} to 10^{15}) are aligned parallel so that the magnetic force within the domain is strong.

When a ferromagnetic material is in the unmagnitized state, the domains are nearly randomly organized and the net magnetic field for the part as a whole is zero. When a magnetizing force is applied, the domains become aligned to produce a strong magnetic field within the part. Iron, nickel, and cobalt are examples of ferromagnetic materials. Components with these materials are commonly inspected using the magnetic particle method.

MAGNETIC DOMAINS

Ferromagnetic materials get their magnetic properties not only because their atoms carry a magnetic moment but also because the material is made up of small regions known as magnetic domains. In each domain, all of the atomic dipoles are coupled together in a preferential direction.

This alignment develops as the material develops its crystalline structure during solidification from the molten state. Magnetic domains can be detected using Magnetic Force Microscopy (MFM) and images of the domains like the one shown below can be constructed.

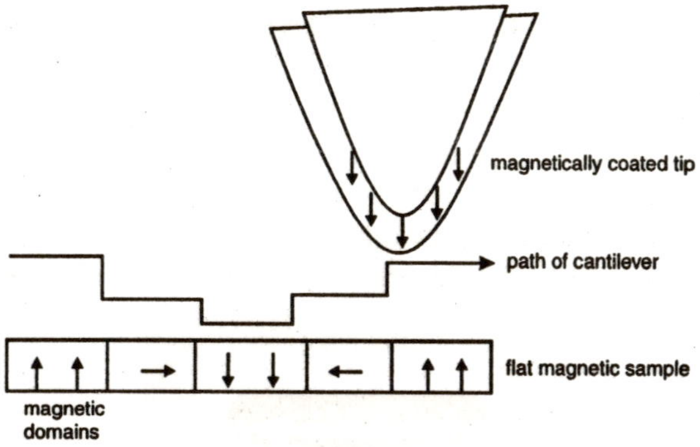

Fig. Magnetic Force Microscopy

During solidification, a trillion or more atom moments are aligned parallel so that the magnetic force within the domain is strong in one direction.

Ferromagnetic materials are said to be characterized by "spontaneous magnetization" since they obtain saturation magnetization in each of the domains without an external magnetic field being applied.

Even though the domains are magnetically saturated, the bulk material may not show any signs of magnetism because the domains develop themselves and are randomly oriented relative to each other.

Ferromagnetic materials become magnetized when the magnetic domains within the material are aligned. This can be done by placing the material in a strong external magnetic field or by passing electrical current through the material. Some or all of the domains can become aligned.

The more domains that are aligned, the stronger the magnetic field in the material. When all of the domains are aligned, the material is said to be magnetically saturated. When a material is magnetically saturated, no additional amount of external magnetization force will cause an increase in its internal level of magnetization.

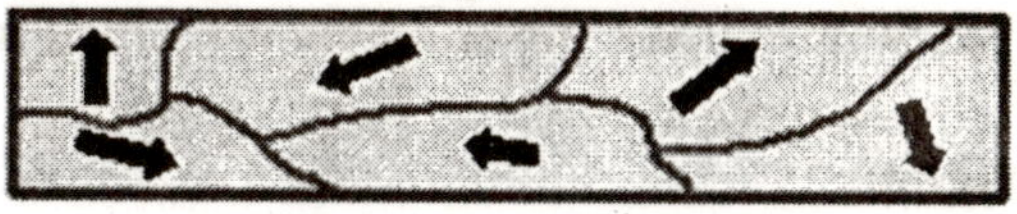

Fig. Unmagnetized Material

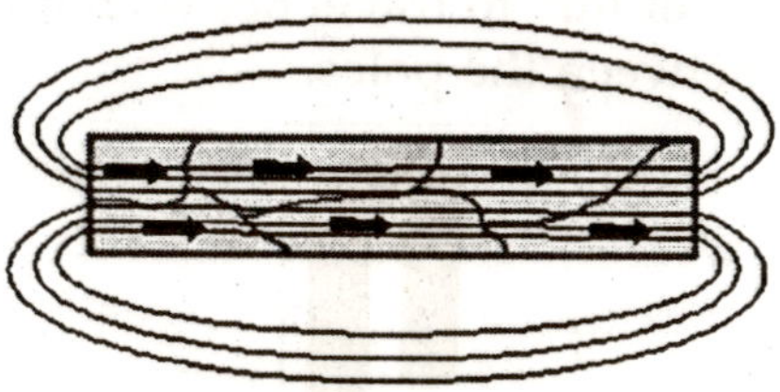

Fig. Magnetized Material

MAGNETIC FIELD CHARACTERISTICS

MAGNETIC FIELD IN AND AROUND A BAR MAGNET

A magnetic field is a change in energy within a volume of space. The magnetic field surrounding a bar magnet can be seen in the magnetograph below. A magnetograph can be created by placing a piece of paper over a magnet and sprinkling the paper with iron filings.

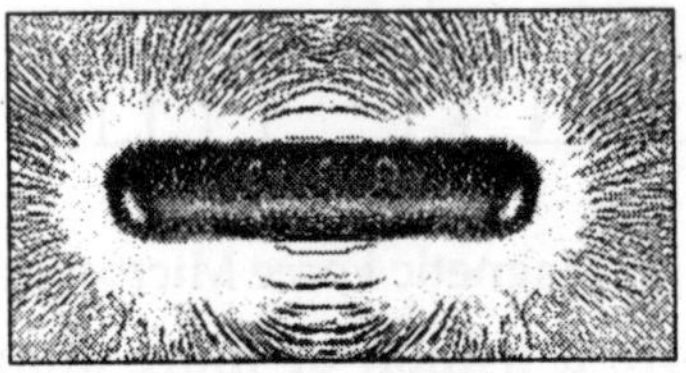

Fig. Bar Magnet

The particles align themselves with the lines of magnetic force produced by the magnet. The magnetic lines of force show where the magnetic field exits the material at one pole and reenters the material at another pole along the length of the magnet. It should be noted that the magnetic lines of force exist in three dimensions but are only seen in two dimensions in the image.

It can be seen in the magnetograph that there are poles all along the length of the magnet but that the poles are concentrated at the ends of the magnet. The area where the exit poles are concentrated is called the magnet's north pole and the area where the entrance poles are concentrated is called the magnet's south pole.

MAGNETIC FIELDS IN AND AROUND HORSESHOE AND RING MAGNETS

Magnets come in a variety of shapes and one of the more common is the horseshoe (U) magnet. The horseshoe magnet has north and south poles just like a bar magnet but the magnet is curved so the poles lie in the same plane. The magnetic lines of force flow from pole to pole just like in the bar magnet.

However, since the poles are located closer together and a more direct path exists for the lines of flux to travel between the poles, the magnetic field is concentrated between the poles.

Fig. Horseshoe and Ring Magnets

If a bar magnet was placed across the end of a horseshoe magnet or if a magnet was formed in the shape of a ring, the lines of magnetic force would not even need to enter the air. The value of such a magnet where the magnetic field is completely contained with the material probably has limited use. However, it is important to understand that the magnetic field can flow in loop within a material.

General Properties of Magnetic Lines of Force

Magnetic lines of force have a number of important properties, which include:

- They seek the path of least resistance between opposite magnetic poles. In a single bar magnet as shown to the right, they attempt to form closed loops from pole to pole.
- They never cross one another.
- They all have the same strength.
- Their density decreases (they spread out) when they move from an area of higher permeability to an area of lower permeability.
- Their density decreases with increasing distance from the poles.
- They are considered to have direction as if flowing, though no actual movement occurs.
- They flow from the south pole to the north pole within a material and north pole to south pole in air.

When a current carrying conductor is formed into a loop or several loops to form a coil, a magnetic field develops that flows through the centre of the loop or coil along its longitudinal axis and circles back around the outside of the loop or coil. The magnetic field circling each loop of wire combines with the fields from the other loops to produce a concentrated field down the centre of the coil.

The magnetic field is essentially uniform down the length of the coil when it is wound tighter. The strength of a coil's magnetic field increases not only with increasing current but also with each loop that is added to the coil. A long, straight coil of wire is called a solenoid and can be used to generate a nearly uniform magnetic field similar to that of a bar magnet. The concentrated magnetic field inside a coil is very useful in magnetizing ferromagnetic materials for inspection using the magnetic particle testing method. Please be aware that the field outside the coil is weak and is not suitable for magnetizing ferromagnetic materials.

QUANTIFYING MAGNETIC PROPERTIES

However, it is necessary to be able to measure and express quantitatively the various characteristics of magnetism. Unfortunately, a number of unit conventions are used. SI units will be used in this material.

The advantage of using SI units is that they are traceable back to an agreed set of four base units–meter, kilogram, second, and Ampere.

Quantity		SI Units (Sommerfeld)	SI Units (Kennelly)	CGS Units (Gaussian)
Field	H	A/m	A/m	oersteds
Flux Density (Magnetic Induction)	B	tesla	tesla	gauss
Flux	ϕ	weber	weber	maxwell
Magnetization	M	A/m	–	erg/Oe-cm³

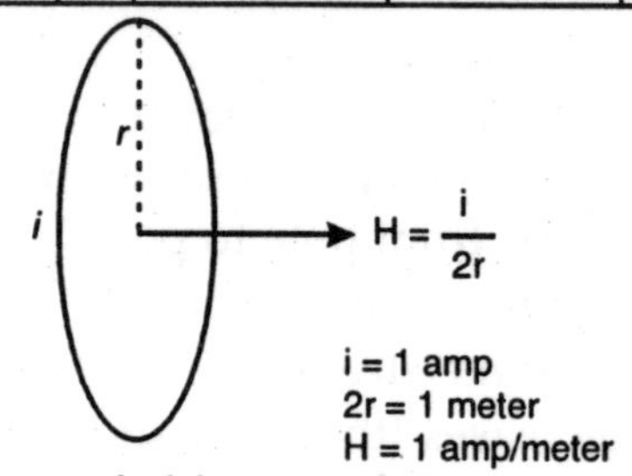

The units for magnetic field strength H are ampere/meter. A magnetic field strength of 1 ampere/meter is produced at the centre of a single circular conductor with a one meter diameter carrying a steady current of 1 ampere. The number of magnetic lines of force cutting through a plane of a given area at a right angle is known as the magnetic flux density, B. The flux density or magnetic induction has the tesla as its unit. One tesla is equal to 1 Newton/(A/m). From these units, it can be seen that the flux density is a measure of the force applied to a particle by the magnetic field.

The Gauss is the CGS unit for flux density and is commonly used by US industry. One gauss represents one line of flux passing through one square centimeter of air oriented 90 degrees to the flux flow. The total number of lines of magnetic force in a material is called magnetic flux, ϕ. The strength of the flux is determined by the number of magnetic domains that are aligned within a material. The total flux is simply the flux density applied over an area. Flux carries the unit of a weber, which is simply a tesla- meter^2. The magnetization is a measure of the extent to which an object is magnetized. It is a measure of the magnetic dipole moment per unit volume of the object. Magnetization carries the same units as a magnetic field: amperes/meter.

Table. Conversion Between CGS and SI Magnetic Units

Quantity	CGS unit	SI units
Magnetic field	$H = 1$ Oe	$H = 1000/4\pi$ A.m^{-1}
Magnetic induction	$B = 1$ gauss	$B = 1 \times 10^{-4}$ Tesla
Magnetization	$M =$ erg.Oe^{-1}.cm^{-3}	$M = 1 \times 103$ A.m^{-1}

THE HYSTERESIS LOOP AND MAGNETIC PROPERTIES

A great deal of information can be learned about the magnetic properties of a material by studying its hysteresis loop. A hysteresis loop shows the relationship between the induced magnetic flux density (*B*) and the magnetizing force (*H*). It is often referred to as the B–H loop. An example hysteresis loop is shown below.

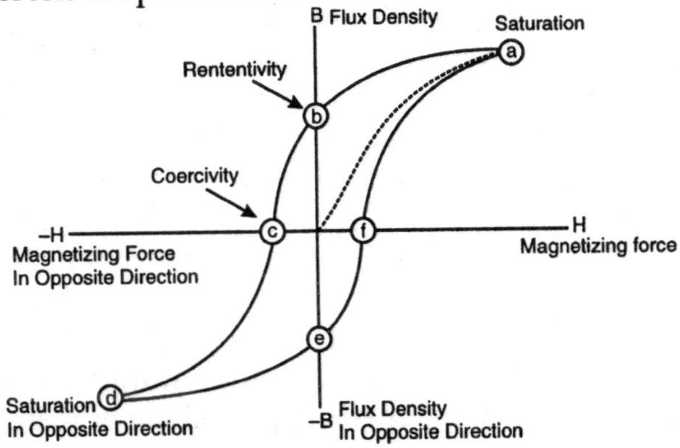

Fig. Hysteresis Loop

The loop is generated by measuring the magnetic flux of a ferromagnetic material while the magnetizing force is changed. A ferromagnetic material that has never been previously magnetized or has been thoroughly demagnetized will follow the dashed line as H is increased. As the line demonstrates, the greater the amount of current applied (H^+), the stronger the magnetic field in the component (B^+). At point "a" almost all of the magnetic domains are aligned and an additional increase in the magnetizing force will produce very little increase in magnetic flux. The material has reached the point of magnetic saturation. When H is reduced to zero, the curve will move from point "a" to point "b." At this point, it can be seen that some magnetic flux remains in the material even though the magnetizing force is zero.

This is referred to as the point of retentivity on the graph and indicates the remanence or level of residual magnetism in the material. As the magnetizing force is reversed, the curve moves to point "c", where the flux has been reduced to zero. This is called the point of coercivity on the curve. The force required to remove the residual magnetism from the material is called the coercive force or coercivity of the material.

As the magnetizing force is increased in the negative direction, the material will again become magnetically saturated but in the opposite direction (point "d"). Reducing H to zero brings the curve to point "e." It will have a level of residual magnetism equal to that achieved in the other direction. Increasing H back in the positive direction will return B to zero.

Notice that the curve did not return to the origin of the graph because some force is required to remove the residual magnetism. The curve will take a different path from point "f" back to the saturation point where it with complete the loop.

From the hysteresis loop, a number of primary magnetic properties of a material can be determined.

- *Retentivity*: A measure of the residual flux density corresponding to the saturation induction of a magnetic material. In other words, it is a material's ability to retain a certain amount of residual magnetic field when the magnetizing force is removed after achieving saturation. (The value of B at point b on the hysteresis curve.)
- *Residual Magnetism or Residual Flux*: The magnetic flux density that remains in a material when the magnetizing force is zero. Note that residual magnetism and retentivity are the same when the material has been magnetized to the saturation point. However, the level of residual magnetism may be lower than the retentivity value when the magnetizing force did not reach the saturation level.
- *Coercive Force*: The amount of reverse magnetic field which must be applied to a magnetic material to make the magnetic flux return to zero. (The value of H at point c on the hysteresis curve.)
- *Permeability, μ*: A property of a material that describes the ease with which a magnetic flux is established in the component.
- *Reluctance*: Is the opposition that a ferromagnetic material shows to the establishment of a magnetic field. Reluctance is analogous to the resistance in an electrical circuit.

PERMEABILITY

As previously mentioned, permeability is a material property that describes the ease with which a magnetic flux is established in a component. It is the ratio of the flux density to the magnetizing force and is represented by the following equation:

$$\mu = B/H$$

It is clear that this equation describes the slope of the curve at any point on the hysteresis loop. The permeability value given in papers and reference materials is usually the maximum permeability or the maximum relative permeability. The maximum permeability is the point where the slope of the B/H curve for the unmagnetized material is the greatest.

This point is often taken as the point where a straight line from the origin is tangent to the B/H curve.The relative permeability is arrived at by taking the ratio of the material's permeability to the permeability in free space (air).

$$\mu(\text{relative}) = \mu(\text{material})/\mu(\text{air})$$

Where: $\mu(\text{air}) = 1.256 \times 10^{-6}$ H/m

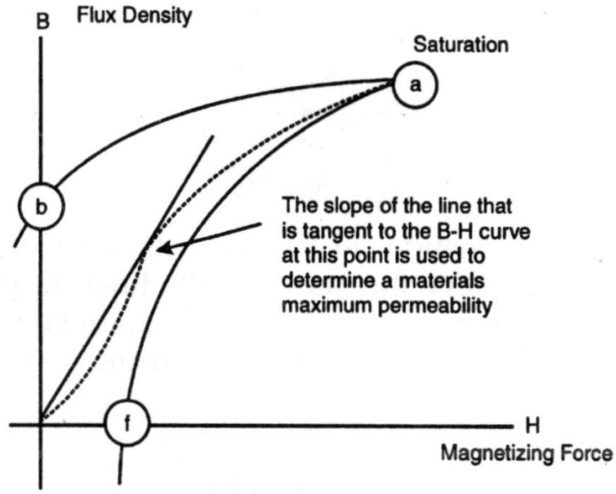

The shape of the hysteresis loop tells a great deal about the material being magnetized. The hysteresis curves of two different materials are shown in the graph.

Relative to other materials, a material with a wider hysteresis loop has:

- Lower Permeability
- Higher Retentivity
- Higher Coercivity
- Higher Reluctance
- Higher Residual Magnetism

Relative to other materials, a material with the narrower hysteresis loop has:

- Higher Permeability
- Lower Retentivity
- Lower Coercivity
- Lower Reluctance
- Lower Residual Magnetism.

In magnetic particle testing, the level of residual magnetism is important. Residual magnetic fields are affected by the permeability, which can be related to the carbon content and alloying of the material. A component with high carbon content will have low permeability and will retain more magnetic flux than a material with low carbon content.

MAGNETIC FIELD ORIENTATION AND FLAW DETECTABILITY

To properly inspect a component for cracks or other defects, it is important to understand that the orientation between the magnetic lines of force and the flaw is very important. There are two general types of magnetic fields that can be established within a component.

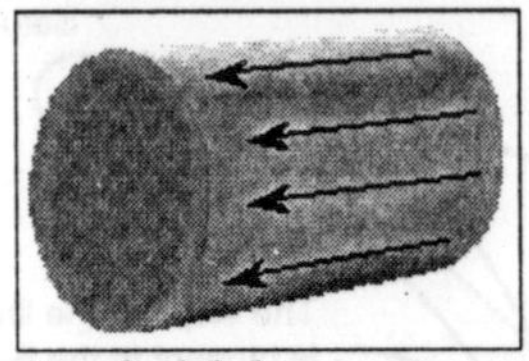

A longitudinal magnetic field has magnetic lines of force that run parallel to the long axis of the part. Longitudinal magnetization of a component can be accomplished using the longitudinal field set up by a coil or solenoid. It can also be accomplished using permanent magnets or electromagnets.

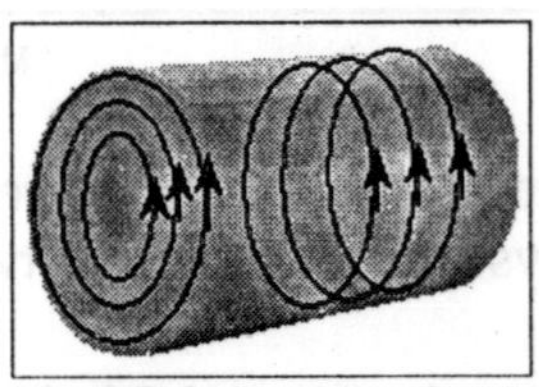

A circular magnetic field has magnetic lines of force that run circumferentially around the perimeter of a part. A circular magnetic field is induced in an article by either passing current through the component or by passing current through a conductor surrounded by the component.

The type of magnetic field established is determined by the method used to magnetize the specimen. Being able to magnetize the part in two directions is important because the best detection of defects occurs when the lines of magnetic force are established at right angles to the longest dimension of the defect.

This orientation creates the largest disruption of the magnetic field within the part and the greatest flux leakage at the surface of the part. An orientation of 45 to 90 degrees between the magnetic field and the defect is necessary to form an indication. Since defects may occur in various and unknown directions, each part is normally magnetized in two directions at right angles to each other.

If the component below is considered, it is known that passing current through the part from end to end will establish a circular magnetic field that will be 90 degrees to the direction of the current. Therefore, defects that have a significant dimension in the direction of the current (longitudinal defects) should be detectable. Alternately, transverse-type defects will not be detectable with circular magnetization.

MAGNETIC FORCES

MAGNETIC FORCE ON A CHARGED PARTICLE

Experimentally, to isolate the nature of the magnetic field, we need to start, as with the electric field, with the simplest thing you can act on, namely a particle. The behaviour of magnetic forces is best indicated by how a charged particle behaves in the presence of such a field. A charged particle moves in the presence of a magnetic field. If the particle stops its motion, the magnetic force on it is zero.

This is true independent of the direction or magnitude of the magnetic field. The mathematical description of the force due to the magnetic field is $F = qv \times B$.This formula describes a force which is, by virtue of the cross-product, always perpendicular to v and B and has a magnitude equal to qvB sinq where q is the angle between v and B.

This is a general result for any charged particle moving in any magnetic field. The circular path seen in the applet results because the force is always perpendicular to the velocity (and also to the B field direction).

If the velocity magnitude and B are constant, then the path of the particle is a circle since the force, and hence the acceleration, of the particle is always perpendicular to its path and constant in magnitude. This fact is used to make two very practical devices of the 20th century: the mass spectrometer and the cyclotron. Both of these have a multitude of scientific and practical uses.

FORCES ON CURRENTS

In the earliest studies of magnetic fields, individual charged particles were not known. The properties of magnetic fields and forces were studied using the effects of fields on current distributions. If we have a bunch of charges flowing through a straight wire of length L, then the net force is

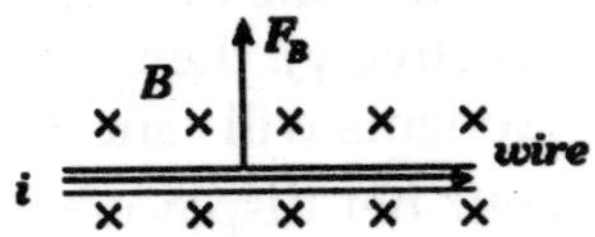

$$\vec{F} = \Sigma_q q\vec{v} \times \vec{B}$$

$$= \Sigma_q \frac{q\vec{L}}{t} \times \vec{B}$$

$$\vec{F} = i\vec{L} \times \vec{B}$$

Force on a wire carrying a current I in a magnetic field B. Where the direction of L is the same as the direction of the current. We can apply the above formulas to the more general case of current distributions of arbitrary shape by breaking any line of current into infinitesimal lengths and integrating over them to find the net force, dF= idL × B

MAGNETIC FORCE ON AN ARBITRARILY SHAPED WIRE

We can generalize the above result to take into account any arbitrary shape of current distribution. For example, consider some arbitrarily shaped wire takes a current i from point a to point b in a magnetic field B which has constant magnitude and direction perpendicular to the plane containing i.

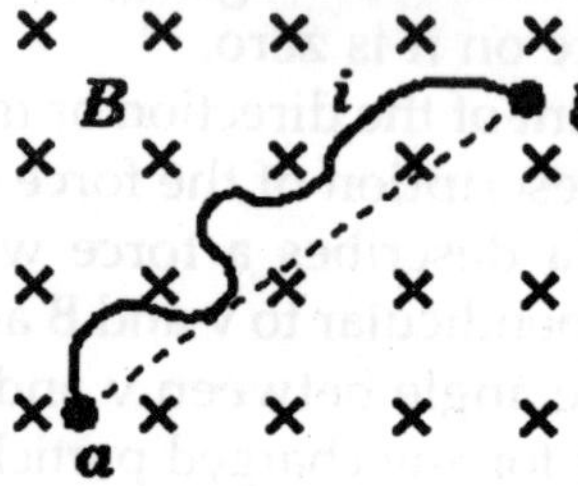

Fig. Current Distribution in an External Magnetic Field

The problem is not as difficult as it might at first appear. Consider any particular piece of the wire.

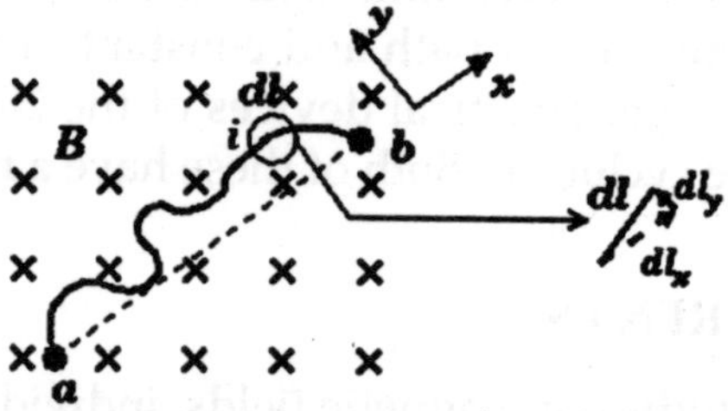

Fig. Blowup of a Section of The Wire

In breaking this distribution into infinitesimal lengths, dl, if we note that we are free to do this such that dl has components that are both perpendicular and parallel to the line between points a and b (we called these the y and x axes, respectively), then note that the contributions on the force due to the y components will cancel.

In going from a to b, our net displacement is only along x. Hence, any +y displacement must eventually be countered with a -y displacement that cancels it. Since the magnetic force on any +y displacement is opposite in direction to that on any -y displacement, the integral over i^*dl_yB must be zero.

Hence, we have a non-zero contribution only from the i^*dl_xB components. We note that the B field is perpendicular to both the x and y components. Hence our result is $F_{net} = iL_{ab}B$ where L_{ab} represents the distance from a to b along a straight line between the two points.

MAGNETIC FIELD OF MOVING CHARGES

In keeping with our previous statement about the association of charges acted upon by fields and the fields produced by those charges, it is only natural to think that, once it is established that magnetic fields act on *moving* electric charges, then moving electric charges should create magnetic fields. In fact, experiment shows just such a relationship! The symmetry inherent in nature between fields and charges is therefore upheld. We find that a single moving point charge q which has constant velocity v in a vacuum produces a magnetic field;

$$B = \mu_0/4\pi \times q\ v \times r/r^2$$

Note the introduction of a new fundamental constant of nature, m_0, the permeability of free space.

We *define* the value of m_0 as being:

$$\mu_0 = 4p \times 10^{-7}\ N{\cdot}s^2/C^2 = 4\lambda \times 10^{-7}\ T{\cdot}m/A$$

So, if m_0 is a constant of nature, why is it defined rather than measured as other constants have been? The reason lies in what was historically recognized as a curious relationship between the permittivity and the permeability of free space. Namely, their product is the inverse of the square of the speed of light, i.e.

$$c^2 = 1/\varepsilon_0\mu_0$$

The reason for this amazing correspondence will be explained later in the semester. For now, we take it that it is useful to define the value of c and hence after measuring ε_0 we define the value of μ_0 to be consistent with the definition of c given the above correspondence.

BIOT-SAVART LAW

In determining the magnetic field for a bunch of moving charged particles which formulate a current, one must be guided by experiment. Starting from measurements of the magnetic field around a long wire carrying a steady current, two French scientists, Jean Biot and Felix Savart, worked out the relationship relating magnetic field strength to the distance from the wire to the point where the field is measured as being 1/R with R being the distance. In keeping with nature's desire for symmetry, there are no preferred directions in space, hence all points on a circle of radius R and centred on the current distribution have the same field magnitude. It was found that the direction of the field is tangent to this circle at each point. Later, it was determined that the strength of the field was proportional to the current I in the wire.

$$B = \frac{\mu_0 I}{2\pi R}$$

Starting from experimental observation of the above, a theoretical formula can be obtained that reflects the contribution due to current through infinitesimal lengths of wire all summed together vectorially through integration. This formula, first worked out by Biot and Savart, is referred to as the Biot-Savart Law is the magnetic field contribution of a current element, i.e. an infinitesimal length dl of current I where dl has the direction of I and r is the vector from the current element to the position where the field is to be calculated.

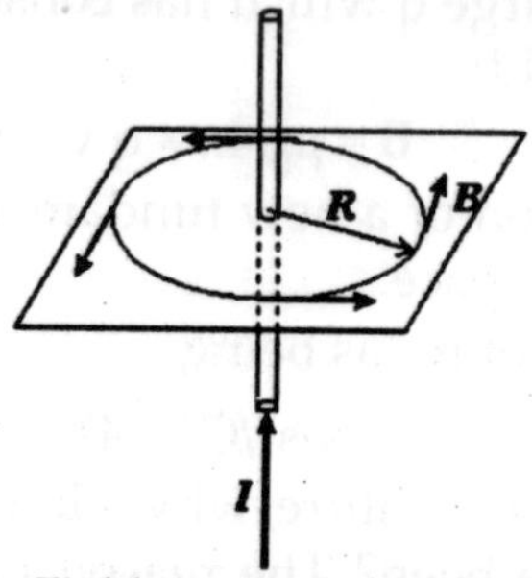

Fig. Magnetic Field Around a Straight Conductor Carr ying a Current I

The Biot-Savart Law is the notion that magnetic fields obey the *principle of superposition*. This means that we can get the net field by just vectorially adding the contributions of the fields due to the infinite number of infinitesimal current elements.

Applying the Biot-Savart Law

To gain experience with applying this law, let's consider some specific examples, starting with the infinite straight wire carrying a steady current I.

Definition of Ampere's Law

Andre Marie Ampere was not in agreement with Biot and Savart as to their experimental determination of the B field. First of all, isolated "current elements" such as appear in the law are difficult to justify "physically". Currents are always part of a complete circuit and hence identifying them as individual, isolated "elements" denies their true nature. In his own experimental and theoretical studies, Ampere determined an alternative law that can be shown to be in agreement with Biot and Savart's Law for the case of an infinitely long, infinitesimally thin wire.

Ampere's Law is stated as:

$$\oint \vec{B} \cdot d\vec{l} = \mu_0 I_{enc}$$

where I_{enc} is the current penetrating the area enclosed by the path over which we integrate. The obvious example to apply this to is the long, thin wire. Here, the wire is carrying current into the page and we wish to find the magnetic field at a point P a distance r away.

As in all examples of using Ampere's Law, we wish to find a closed path, i.e. one that bounds an area, that allows us to separate B from the integral. To do this, we need a path in which B is constant in magnitude and parallel, antiparallel, or perpendicular to d*l* for every part of the path. The reasons for this are:

- In order to be mathematically sensible, the path must define an area through which the current passes, i.e. a non-closed path does not define a definitive area through which we can unambiguously state what current is contributing to the B field along the path. Therefore, the path must be closed.
- Parts of the path must be parallel (or anti-parallel) to the B field so that we have something to evaluate for B·d*l*. If, for part of the path, we have B perpendicular to d*l*, then, for those parts, B·d*l* = 0. In the end, we wish to know the magnitude of B. Therefore, we must have B·d*l* be known over the whole path and known in such a way that *all* of B is known, not just a component of it.
- Clearly, for this to work, you can only use Ampere's Law for cases in which you already know a lot about the behaviour of B. You need to know its direction and something about how its magnitude changes over the possible paths you might choose. In addition, the B field must have enough symmetry to make it possible for you to complete steps 1 and 2 above.

Applying Ampere's Law to a Long, Thin Wire

The obvious path for the long, thin wire is shown below as the dashed line labeled d*l*.

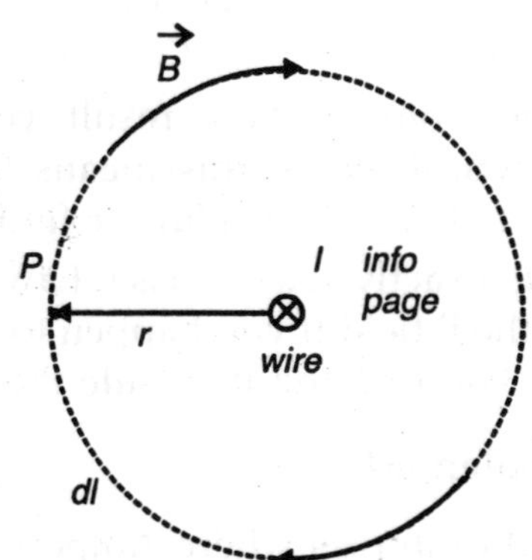

We use this circular path of radius r since we know, from the Biot-Savart Law, that the magnetic field around the wire is tangent to the circle in the clockwise direction and that it's magnitude depends only on r, not on position parallel to the wire.

Thus, we use Ampere's Law to derive:

$$B.dl=\mu_0 I_{enc}, \; B=\mu_0 I/2\pi r$$

To be sure that Ampere's Law is consistent with the Biot-Savart Law, let's see if it works for some path other than the circular one we chose.

For example, look at the path below which has two circular arc sections centred on the wire carrying current into the page. One arc has radius a and the other radius b. They are connected by straight sections.

Note that the path is closed and that we do know the characteristics of the B field. The circled numbers indicate that we have identified 4 pieces of the path for the application of Ampere's Law.

These four pieces give where we need to know what B_1 and B_3 are and we have used the fact that, for the straight paths 2 and 4, the B field is perpendicular to the path *dl* at every point along the path. The Biot-Savart Law also tells us that

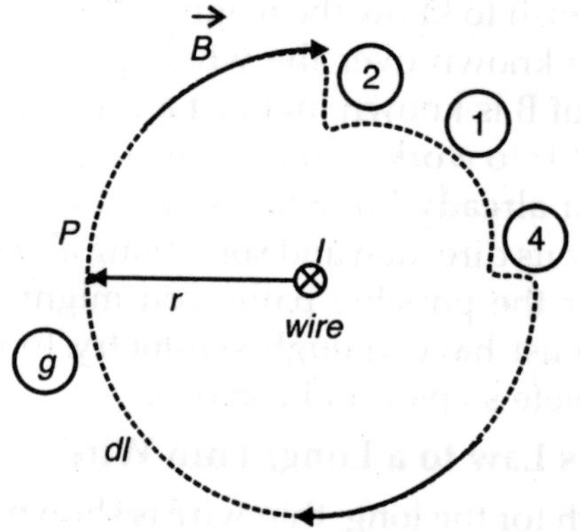

Fig. Ampere's Law Applied to a Straight Current into the Page and an Amperian path which is not Circular $B = \mu_0 I/2\pi b$

If you plug these into our previous result, you find that Ampere's Law is verified. In practical terms, this means that Ampere's Law is always valid. It may not always be useful for finding the B field.

In this respect, it is exactly like Gauss's Law in that you can only apply it to learn about the B field if you happen to recognize a symmetry of the field that allows you to bring B outside the integral of B·*dl*.

Magnetic Field of a Solenoid

We can readily find examples where Ampere's Law is more useful, by virtue of being easier to apply, than the Biot-Savart Law. For example, as was the case for the electric field, it was deemed to be very practical to have a device which could "store" magnetic field as a capacitor "stores" an electric field.

We know that what the capacitor really stores is electric charge separation, but one of the properties for which a parallel-plate capacitor is useful is its ability to produce a uniform electric field between its plates.

The equivalent device for magnetic fields is the solenoid. It's simply a conducting wire wrapped into a cylindrical shape. Even though the actual shape of the wire is helical, for densely packed wrapping we can actually consider the solenoid to be a bunch of closely spaced coils. Near the centre of each coil we know that the magnetic field is nearly perpendicular to the plane of the coil.

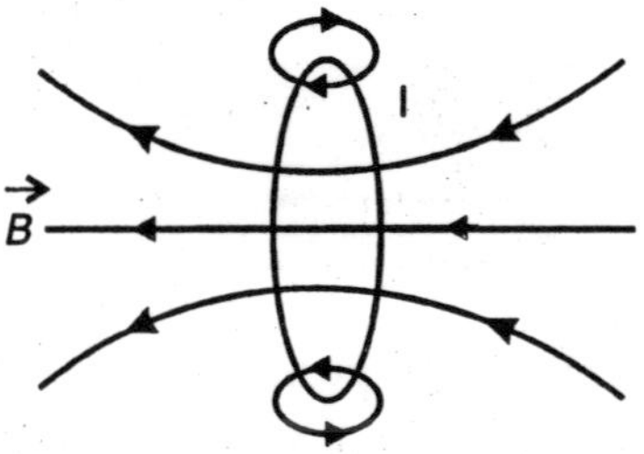

Fig. The Magnetic Field Near the Centre of a Single Coil Carrying a Current I

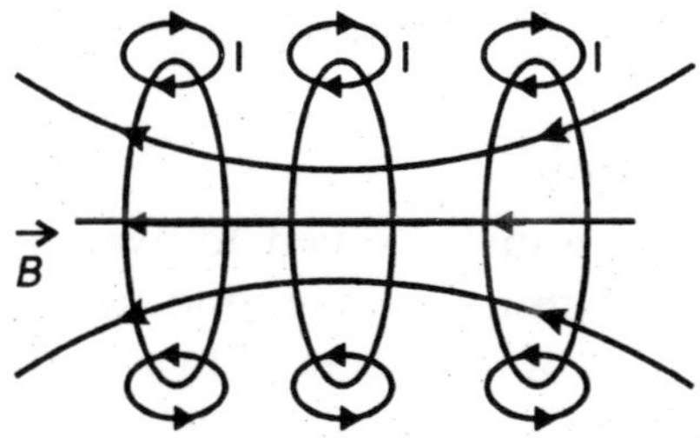

Fig. The Magnetic Field Near the Centre of a Set of three Coils all Carrying a Current I

If we place coils on either side of the first and let them all carry the same current I in the same direction, then the magnetic field lines will link together near the centre to make a field that is approximately uniform in direction and strength near the centre of the set of coils. Placing many coils in close proximity to each other yields the solenoid field.

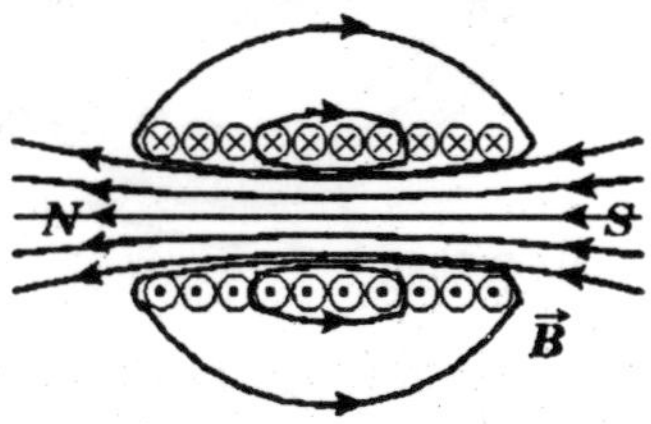

A solenoid approximates many current-carrying coils spaced much more closely than the coil diameter.

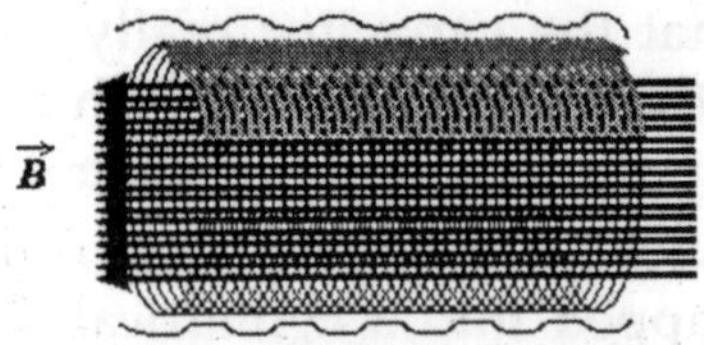

A perfect solenoid has coils so close together that the magnetic field is zero outside the solenoid and perfectly uniform inside. Approximating the field as constant in direction and magnitude near the centre of the solenoid allows us to use Ampere's Law to calculate its magnitude.

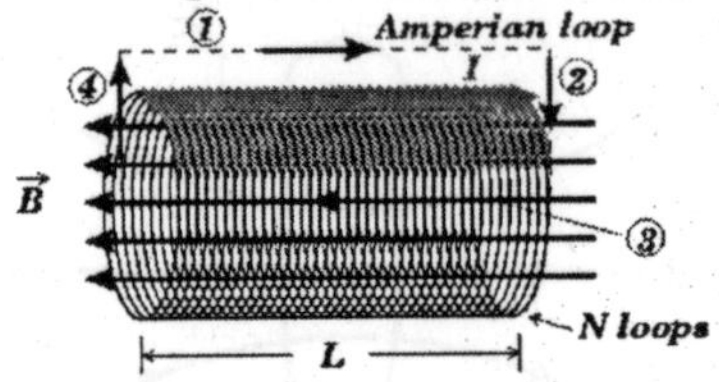

Fig. Applying Ampere's Law to a solenoid

Choosing a flat rectangle as the Amperian loop that the contributions to the loop integral can be broken into four parts. The area bounded by the flat rectangle is penetrated by N loops over a length L.Hence our result is *B.dl*= $B_{solenoid}L = m_0NI$ $B_{solenoid} = m_0nI$.where n = N/L is the linear density of current loops.

Magnetic Field of a Toroid

We can use what we just derived to examine the case of another device that produces a uniform magnetic field. In this case, we take advantage of our analysis of the solenoid to ask what happens if we bend a solenoid into a circle so that the ends join.

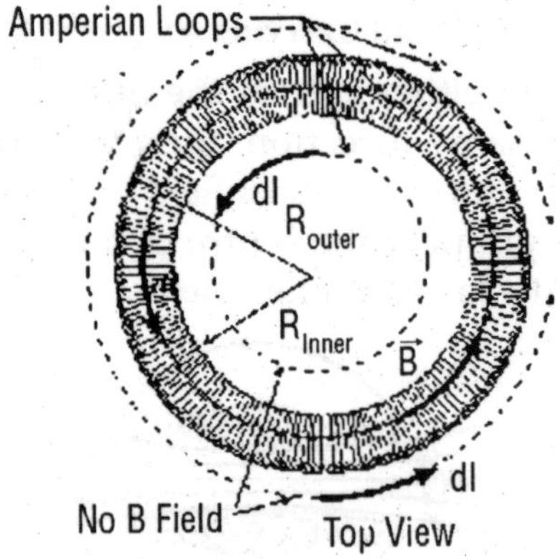

The new configuration still has approximately zero field in the regions outside the volume contained by the coils, but the field inside that volume is again approximately uniform if the distance between the coils is small compared to the size of the coils. The above device is a toroid. Ampere's Law applied to a toroid.

Note that Amperian loops which lie wholly outside the volume contained by the toroidal coils experience no magnetic field. We expect the magnetic field to have circular symmetry about the centre of the toroid using the same reasoning as for the solenoid. Hence, we expect it to be most useful to use circular paths for evaluating the Ampere integral of magnetic field and path.

For any circular path whose area is not intersected by the coils, the magnetic field is zero and the current penetrating the area is, by definition, zero. Hence, $B_{r < Rinner} = 0$. For a circular path whose radius is greater than R_{outer}, we also expect to have no magnetic field. This might seem to be a contradiction with Ampere's Law because the current in the toroid coils definitely penetrate the area contained, however, note that the coils take current through the area (out of the page) at R_{inner} and *back into* the page at R_{outer}. So, the net current penetrating the area contained by the circular loop is zero since every loop of the wire carries current through the area twice-in opposite directions! So it is consistent that the magnetic field be zero.

Magnetic Field of an Infinite Sheet of Current

We can consider an instance of a current distribution which does not have cylindrical symmetry, but which is susceptible to Ampere's Law for finding the magnitude of the magnetic field. Consider a sheet of current which is infinitesimally thin but infinitely long and wide. The sheet has a linear current density (i.e. current per unit length), λ.

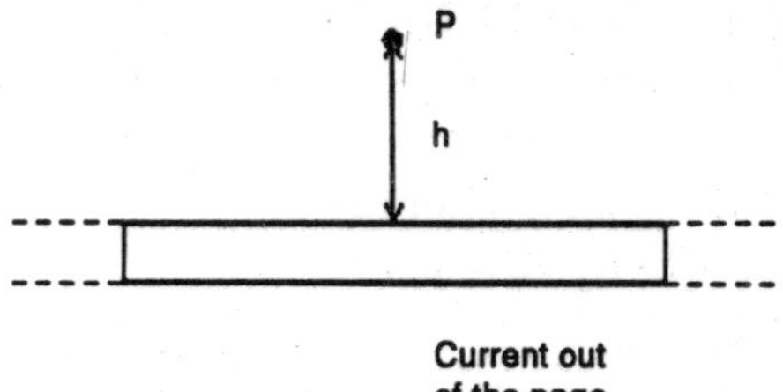

Fig. An Infinite Sheet of Current with Current Per Unit Length λ. We Wish to Find the Magnetic Field Direction and Magnitude at Point P a Distance h Away from the Sheet.

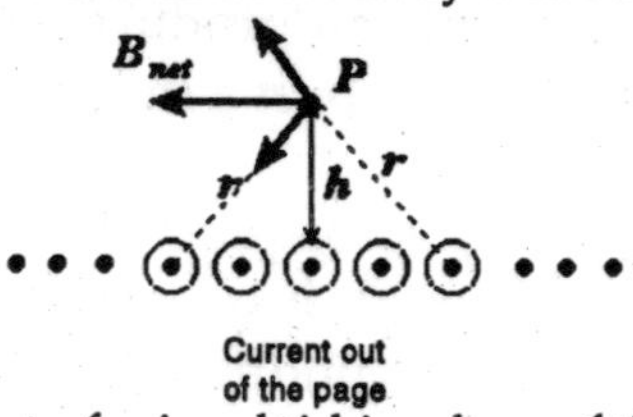

Fig. Consider a set of wires laid in place of the Current Sheet

We can replace the sheet with an infinite set of wires arranged so that each wire carries current consistent with an overall current per unit length of l as in the case of the current sheet. The wires which are equidistant from the line from the set of wires to point P have magnetic fields whose vertical components cancel and whose horizontal components add.

Hence, the net magnetic field is in the horizontal direction. This field is uniform since the distribution of wires is infinite, i.e. any position for point P can be considered as the "middle" of an infinitely long current sheet.

Note also that the magnetic field direction for the infinite sheet is independent of the distance of P from the sheet and that the same arguments for extending consideration of a finite set of wires to an infinite current sheet state that the magnetic field direction on the other side of the sheet has the field pointing in the opposite direction.

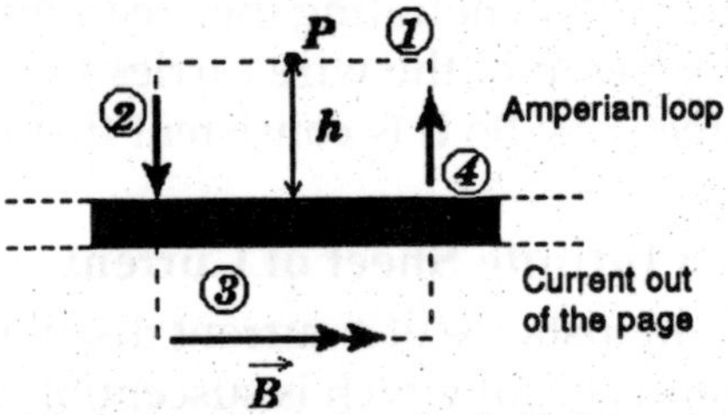

Fig. The Amperian loop for the infinite current sheet

The loop gives us $B.dl = \mu_0 I_{enc.},$ $B_1 \cdot dl + B_2 \cdot dl + B_3 \cdot dl + B_4 \cdot dl = m_0 \lambda L$. Notice that the magnetic field is always perpendicular to parts 2 and 4 of the path. Part 3 is along the field direction below the sheet since, as stated before, the same symmetry arguments work here to say that the field should be parallel to the sheet, but in the opposite direction to the field on the other side of the sheet. Therefore,

$$B_1 \cdot dl + B_2 \cdot dl + B_3 \cdot dl + B_4 \cdot dl = \mu_0 \lambda L$$

$$B_P L + 0 + B_P L + 0 = \lambda_0 \lambda L$$

$$B_P = \tfrac{1}{2} \mu_0 \lambda$$

FARADAY'S LAW

Symmetry in nature predicts that, if a magnetic field acting on a loop of conductor carrying a current can produce a torque on the loop then a torque acting on a conducting loop in a magnetic field should produce a current.

Symmetry in producing a torque for a loop with an externally maintained current going through it is balanced in nature by having an externally applied torque on a conducting loop produced a current in the opposite direction.

Note that the symmetry cannot be "perfect" in that the direction of the current that is induced in the loop due to the externally maintained torque must be opposite to the direction of current that would create a torque in that direction.

This is necessary due to energy conservation. If the effect happened as in our first case then more current would go through the loop and the induced torque would increase.

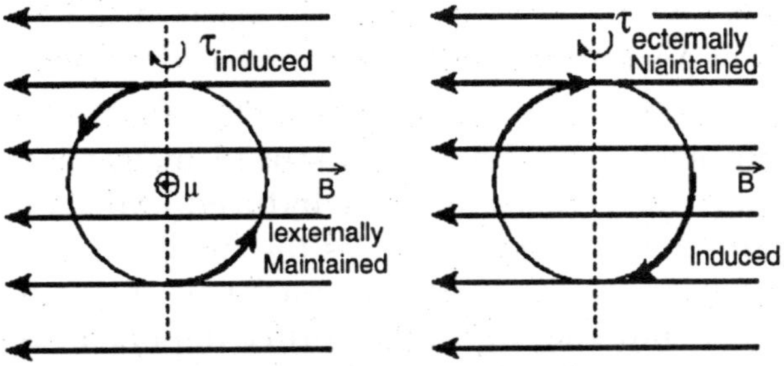

This violates energy conservation as the change in angular position due to the external torque would induce still more change in angular position. Hence we note that if the effect of induced current does take place, the current must be induced in a direction that *opposes* the external torque that causes it.

Faraday was the first to show that this induced current effect does happen. In studying the effect, we find that the magnitude and direction of the induced current effect is due to a change in the magnetic flux through an area in space. This effect is referred to as electromagnetic induction and the quantitative description of the effect is Faraday's Law. Specifically, Faraday's Law states that $\varepsilon = -d\phi_B/dt$.

In other words, Faraday expressed his effect by saying that the magnitude of the EMF induced by any change of magnetic flux is proportional to the magnitude of the change in magnetic flux. The induced EMF is what causes the current to flow.

To correspond to the necessity of energy conservation, the direction of the EMF must be such that the induced current would oppose the change in the magnetic flux. This was best expressed by H.F.E. Lenz in Lenz's Law.

The Direction of any Magnetic Induction Effect must Oppose the Cause of the Effect

The easiest way to generate a static picture of this is to observe the four cases below of a bar magnet being pushed toward and away from a conducting loop. The current created by the induced EMF must oppose the change in each case.

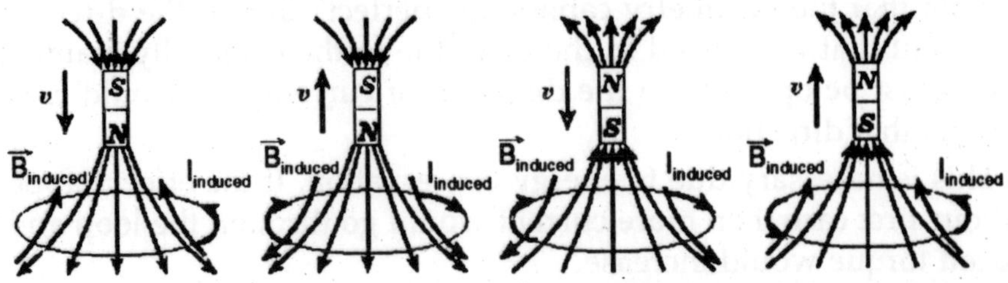

Fig. A Stationary Conducting Loop with a Bar Magnet that can be Alternately Pushed toward or Away from the Loop with Either of two Orientations of the Poles (i.e. North pole Toward the Loop or Away from it)

It shows, in each case, how the induced magnetic field from the induced current (which comes from the induced EMF assuming the loop has some amount of resistance) opposes the change in magnetic flux.

Since magnetic flux is a scalar, it might be difficult to think of a "direction" for it, but the sign of the flux depends on the arbitrary convention for the sign for area, A. Once that is set, then the dot product of B and A gives the positive or negative sign for the flux. The negative sign in Faraday's Law then gives the appropriate sign for the direction of the induced EMF.

The induced current creates an induced field which points opposite to the external magnetic field from the bar magnet since the flux going down through the loop is increasing with time.

For the third case, the induced current creates an induced field which points opposite to the external magnetic field since the flux pointing up through the loop is increasing. Verify that the second and fourth cases shown make sense to you following the same argument. The importance of Faraday's effect is hard to overestimate. Most of the world runs on electricity generated as a result of it.

Faraday's Law Applied

If we consider the case for a conducting bar sliding over rails in the presence of a magnetic field, we have a good case for examining both Lenz's Law and considering a quantitative example of Faraday's Law.

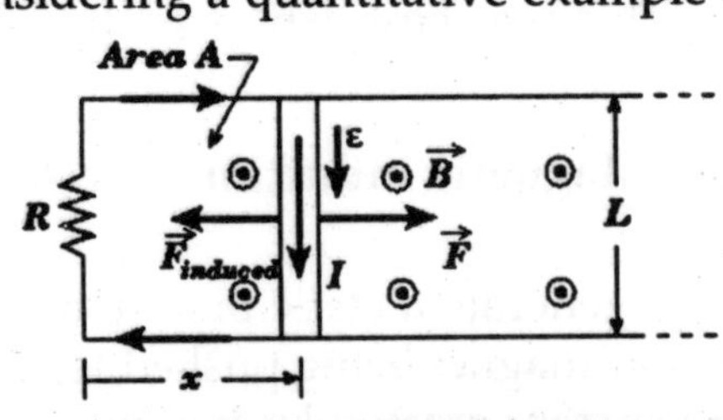

Fig. A Conducting Bar Slides on Frictionless Conducting Rails in the Presence of a Magnetic Field.

MOTIONAL ELECTROMOTIVE FORCE

We consider conducting material to have a number (essentially infinite) number of electric charge carriers that are free to move (albeit with resistance unless the conductor is a perfect conductor) throughout the material in any way demanded by external electric and magnetic fields.

That means we can consider the source of an induced EMF by looking at the direct effects of an external magnetic field on charges in a moving conductor.

For example, if we consider a conducting bar of length L moving in a plane perpendicular to a constant magnetic field.

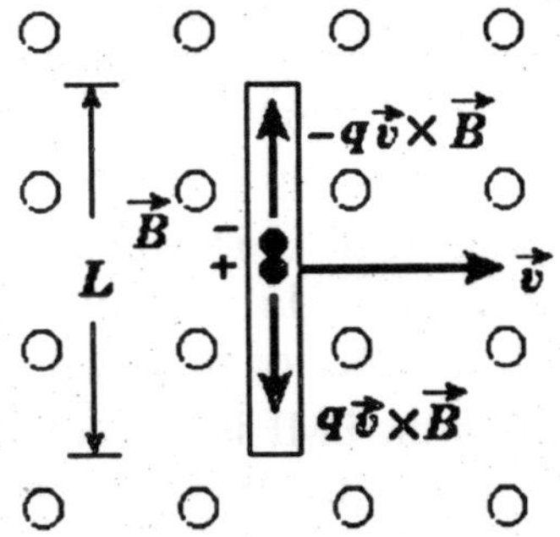

Fig. A Conducting Bar Moving in a Magnetic Field Experiences Separation of Positive and Negative Charges within the Bar.

The positive charges are pushed to the bottom by the velocity cross magnetic field force and the negative charges are pushed in the opposite direction.

The charge separation creates an electric field in the downward direction. This electric field builds as the charge separation grows until the force on the charges due to the resulting electric field cancels the force due to motion in the magnetic field.

That means that we have, if we assume the charge q to mean positive charge,

$$e_{induced} = vBL$$

where the motional EMF, ε, is just the average electric field divided by the length of the bar. It is induced by the motion of the bar in the magnetic field.

The magnitude of this EMF is equal in magnitude to the product of the velocity, the magnetic field magnitude, and the length of the bar.

If we attach this bar to frictionless rails which provide a resistance R and provide an external force on the bar to maintain its constant velocity, then the bar is a *source* of EMF and just like any other EMF source, can drive a current through the circuit.

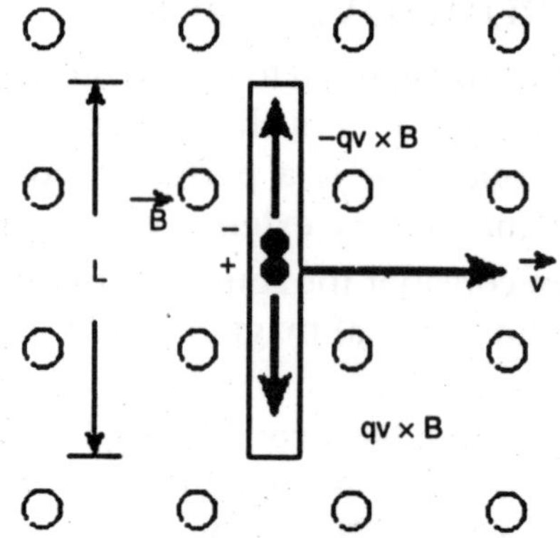

Fig. A Conducting Bar Moving on Frictionless Rails Acts as an EMF Source for Pushing Current Through the Circuit Formed by the Bar and the Rails.

Note that the motional EMF is identical to the induced EMF we calculated through Faraday's Law. In fact, motional EMF is one example of Faraday's Law. It has a number of practical applications.

Note that the direction of the EMF in both the top and bottom pieces of the loop point to the right. Hence the bottom EMF pushes current counter-clockwise about the loop and the top EMF pushes current clockwise.

The motional EMF on each of the four sides of the conducting loop can be calculated. The net EMF pushes current counter-clockwise around the loop.

The net EMF for the circuit is therefore $e_{net} = e_{bot} - e_{top} = B_0 l v_y (y_{bot} - y_{top}) = B_0 l w v_y$ if we arbitrarily assign counter-clockwise as the positive direction. The induced current in the loop is $I_{induced} = e_{net}/R = B_0 l w v_y/R$.

INDUCED ELECTRIC FIELDS

In general, since the force on charge carriers in a conductor is given by F = qv × B,we can calculate the motional EMF for a conductor moving in a magnetic field via the following formulation.

$$E \cdot dl = \varepsilon = -\, dF_B/dt$$

In general, we can calculate motional EMF's, i.e. the EMF generated in *moving conductors* through the interaction of the motion and external magnetic field. The second equation shows that Faraday's Law predicts the creation of an induced electric field in the presence of a time-varying magnetic flux.

Note that this second equation and the original expression of Faraday's Law are equivalent and *always* valid. The calculation of motional EMF's is consistent with Faraday's Law *only* for the case of motion of a conductor in a magnetic field.

AMPERE-MAXWELL LAW

James Clerk Maxwell was responsible for setting down our current understanding of the fundamental importance of the equations that make up Gauss's Law, Faraday's Law, and Ampere's Law. In deciding on a suitable description of electrodynamics, he sought to place the mathematical description on the same footing as Newton's Laws of Motions.

That he succeeded is testimony to his mathematical skill and great physical insight. To start, we should review here the form of the laws as we currently describe them.

Gauss's Law for Static Electric Charges	$E \cdot dA = q/\varepsilon_0$
Gauss's Law for Magnetism	$B \cdot dA = 0$
Ampere's Law	$B \cdot dl = \mu_0 I_{enc.}$
Faraday's Law	$E \cdot dl = -d\phi_B/dt$

where the second equation represents the fact that there are no magnetic charges observed (so far) in the universe, hence there are no point-like sources or sinks of the magnetic field. We first note the degree of mathematical symmetry present in the equations despite the differences in behaviour of electric and magnetic fields.

It's also clear that this symmetry between electric and magnetic fields would be perfect except for two obvious asymmetries within the equations themselves. The first is that there are no magnetic charges as there are electric charges (something that has motivated physicists to seek out such charges for decades) and the second is that, electric fields can be created in *two* ways: either from static electric charges (Gauss's Law) or induced by time-varying magnetic fields (magnetic fluxes to be exact), but there is no corresponding possibility of creating magnetic fields from either static magnetic charges (they have not been found to exist) nor from time-varying electric fields.

If we simply assumed that the latter symmetry had to exist, why would it not be apparent in experiments that were done up to the 1800's? In this, we can make a small digression that points out one of the prime means of gaining insights into physics research.Let us infer that nature is respectful of symmetry mathematically. What would Maxwell's Equations look like completely symmetrically?

First, the existence of magnetic charges would change the form of Gauss's Law applied to static charges. We would have $E \cdot dA = q_E/\varepsilon_0$, EB $\cdot\, dA = \mu_0\, q_B$.. If units are to be kept consistent. What about currents involving electric and magnetic charges? Certainly there would be contributions to both Ampere's Law and Faraday's Law.

These contributions, again looking at consistent units, would look as follows: $E \cdot dl = -d\phi_B/dt\ \mu_0 I_{B.}$ where the subscripts E and B depict electric or magnetic charges.

The k in the revamped Ampere's Law is a placeholder for the correct proportionality constant for this term. We can find the value of this constant in terms of the fundamental constants m_0 and e_0 purely from dimensional analysis.

Since magnetic monopoles do not exist, we deal simply with the additions necessary to consider induced magnetic fields from time-varying electric field fluxes.

This would affect only Ampere's Law and the term added would assume that a time-varying electric flux would induce a B field. To make sure that the units are correct, we would want to have Ampere's Law extended as follows:

$B \cdot dl = \mu_0(I_{enc.} + I_{displacement})$,

where we want $I_{displacement} = d\phi_{E/}dt$,

That is, this new term $I_{displacement}$ is proportional to the time-rate of change of electric flux. The simplest form of converting electric flux (SI units of $N/(C \cdot m^2 \cdot s)$) to current (C/s) is, interestingly enough, multiplication by a constant with the units $C^2/(N \cdot m^2)$. Note that these are precisely the units of the permittivity constant, $\varepsilon_{0.}$ Hence, $I_{displacment} = \varepsilon_0\ d\Phi_E/dt$.

This comes directly from dimensional analysis and a "belief" in the need for symmetry among the equations for electricity and magnetism.

Note that this ansatz also gives rise to an assumption as to why the effect might be hard to observe. Note that, if the above form is correct, then Ampere's Law is changed to have the form $B \cdot dl = \mu_0 I_{enc.} + \mu_0 \varepsilon_0\ d\Phi_E/dt$.

This means that the time-varying electric flux term is multiplied by *two* small terms rather than just one. This might make the effect weak enough to be masked by other experimental interferences unless one is very careful.

In fact, the form we have derived for $I_{displacement}$, referred to as the displacement current, does exist and can be verified experimentally. It was motivated by Maxwell theoretically not from the symmetry and dimensional analysis argument above, but from the following consideration of a weakness in Ampere's mathematical description of his law.

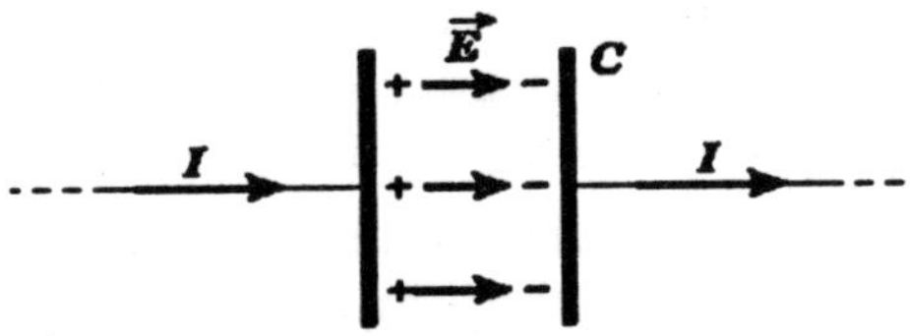

Fig. A Charging Capacitor and the Corresponding Time-varying Electric Field Between its Plates

Consider the case of a capacitor connected to conducting lines. If one of the lines carries a current I, then the other line connected to the other plate must also carry such a current due to the electric field of the previous plate.

If we apply Ampere's Law to a circular disk around the wire, then the magnetic field is expected to be constant in magnitude and to have a direction which follows the boundary. The magnitude of the B field is proportional to the current penetrating the area bound by the path.

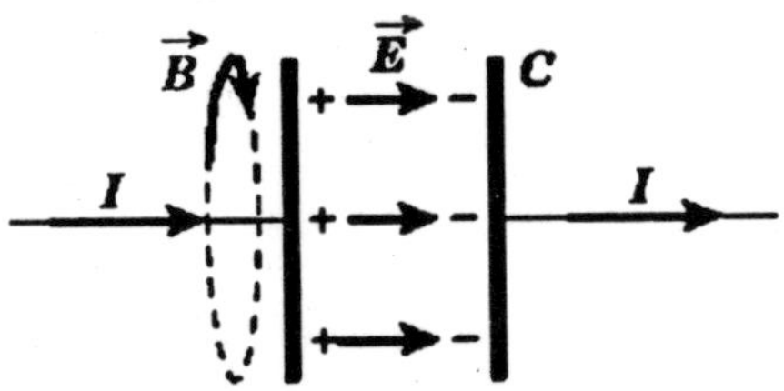

Fig. Magnetic field Around a current as it Approaches the Capacitor

Unfortunately, the mathematical description of the area bounded by a curve is rather loose.

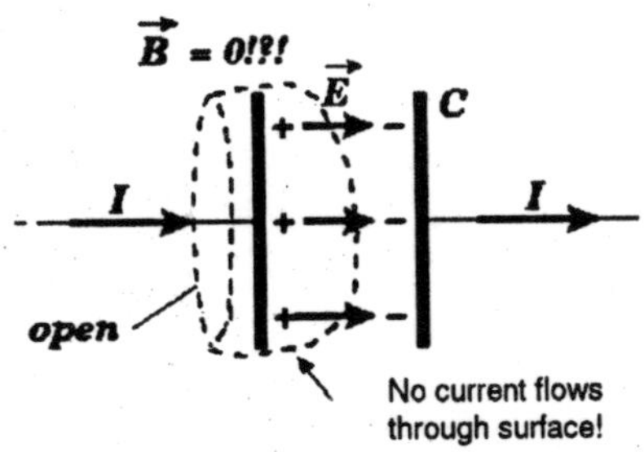

Fig. Magnetic field around the same perimeter which encloses a different area

Ampere's Law, as stated, is a line integral which can be shown to be equivalent for the two areas and in fact for any area bounded by the same circular curve around the wire in the region outside the capacitor.The unfortunate part, though, is that no current penetrates the second surface and hence the prediction is that the magnetic field should be zero around the boundary.To resolve the obvious inconsistency, Maxwell generalized Ampere's Law to include the displacement current matching just the characteristics we gave on the basis of the dimensional argument above.

To verify the form of the displacement current, note that a parallel-plate capacitor builds charge according to the current impinging on the left plate for our figure. This current is related to the change in electric flux as follows:

$$\Phi_E/dt = I/\varepsilon_0.$$

If we multiply this by e_0, we see that we get the displacement current as being equal to the current in the wires attached to the plates of the capacitors, hence we expect to get the same value for the magnetic field no matter whether we draw the surface so as to have the current in the wire penetrate the surface or to have the time-varying electric field between the capacitor plates penetrating the surface.

Chapter 9

Wave Generation and Propagation

To perform realistic simulations of green water and bow impact loading, the method has to be extended with wave generation options. There are different ways to create a wave field in a computational programme:

- Prescribe a wave at the inflow boundary using theory, for example Stokes theory or a superposition of linear waves.
- Use a wave maker, which can be modelled as a moving object.
- Use another calculation method, which calculates kinematics away from an object that will be prescribed at the domain boundaries.

The waves considered are long crested waves, so the incoming wave is two-dimensional of shape. Of course, when an object is present, the wave will be disturbed and three-dimensional aspects become important. At the boundary of the domains appropriate conditions are needed to let the wave travel in and out of the domain in an undisturbed way. Therefore, at the inflow boundary fluid wave kinematics will be prescribed according to wave description theories.

The undisturbed wave will be prescribed, so the disturbance due to wave diffraction on an object is not taken into account. At the outflow boundary conditions should be imposed, such that the wave can leave the domain undisturbedly. Determining these conditions is difficult, because no information about the wave is present near the outflow boundary. After describing the wave theories and conditions at the outflow boundary, attention will be paid to possible energy dissipation due to the upwind discretisation of the convective terms.

Also the influence of the treatment of the free surface on wave propagation, like the choice for free surface velocities and the displacement algorithm, is investigated. To validate the numerical model, simulations have been performed, which can be checked against theory or experiments. First, results will be shown of two-dimensional wave simulations without the presence of an object. Further, a spar platform has been put into the flow.

Forces and wave heights have been calculated, which can be compared with experimental results. Finally, as the most demanding test case, green water on the bow of a moving FPSO is simulated.

WAVE

At the inflow boundary the wave will be initiated using a theoretical wave description. In this thesis the wave is usually travelling in positive x-direction, making the left domain wall the inflow boundary.

The wave is prescribed as head wave, but of course by rotating the body in the domain every wave direction can be achieved. For the simulation of a wave in the domain, the velocities and wave height given by wave theory are prescribed.

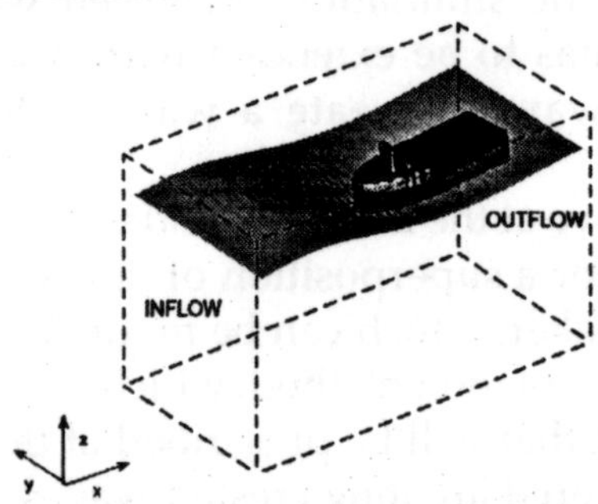

Fig. The Wave Travels in Positive Xdirection.

The waves that can be formed in COM FLOW are based on Airy wave theory describing linear waves, and 5th order Stokes theory describing nonlinear waves. Also a superposition of linear wave components can be used to generate an irregular wave.

Airy Wave Theory

The theory for the generation of a linear wave can be derived in the following way. Mass conservation in a volume filled with an incompressible fluid leads to the continuity equation

$$\nabla . u = 0,$$

with u = (u,w) the velocity vector in two dimensions. The velocity potential F is defined by the equations

$$\frac{\partial \Phi}{\partial x} = u, \frac{\partial \Phi}{\partial z} = w,$$

with × and z the horizontal and vertical coordinate, respectively. Substituting F into the continuity equation leads to a Laplace equation for the velocity potential.

$$\frac{\partial^2 \Phi}{\partial x^2} + \frac{\partial^2 \Phi}{\partial z^2} = 0.$$

When assuming the water surface slope very small, the potential is written as a sine with an amplitude depending on the water depth

$$\Phi(x,z,t) = P(z)\sin(\omega t - kx + \phi),$$

where t is the time, k the wave number, and ϕ the phase angle. P(z) can be determined by substituting Equation into the Laplace equation and solving the differential equation for P(z), which finally results in

$$\Phi(x,z,t) = (C_1 e^{kz} + C_2 e^{-k2})\sin(\omega t - kx + \phi),$$

The constants C1 and C2 can be determined using the boundary conditions at the sea bed and the free surface. At the sea bed a no-leak condition holds, which is given by

$$\frac{\partial \Phi}{\partial z} = 0 \quad \text{at } z = -h.$$

Substituting this boundary condition into Equation reduces the two unknowns to one:

$$\Phi(x,z,t) = C\cosh(k(z+h)\sin(\omega t - kx + \phi).$$

The unknown constant C can be determined by using the free surface dynamic boundary condition that is deduced from the Bernoulli equation, taking into account the small wave steepness.

The condition at the free surface is linearised around the calm water level (z = 0), which leads to:

$$\frac{\partial \Phi}{\partial t} + g\zeta = 0 \quad \text{for } z = 0.$$

The free surface elevation ζ can be derived from this equation after substitution of the velocity potential, resulting in

$$\zeta(x,t) = \zeta_a \cos(\omega t - kx) \ \text{ with } \ \zeta_a = -\frac{\omega}{g} C\cosh(kh).$$

Rewriting the expression for ζ_a, the constant C can be determined as:

$$C = -\frac{\zeta_a g}{\omega} - \frac{1}{\cosh(kh)}.$$

Substituting this into Eq leads to the final expression for the velocity potential

$$\Phi(x,z,t) = -\frac{\zeta_a g}{\omega}\frac{\cosh(k(z+h))}{\cosh(kh)}\sin(\omega t - kx + \phi).$$

The linearised free surface kinematic boundary condition given by

$$\frac{\partial z}{\partial t} + \frac{1}{g}\frac{\partial^2 \Phi}{\partial t^2} = 0 \text{ for } z = 0$$

leads to the dispersion relation, which connects the wave number and frequency in the following way:

$$k \tanh(kh) = \frac{\omega^2}{g}.$$

For deep water, the dispersion relation reduces to $k = \omega^2/g$.

$$u(t,x,z) = \frac{\zeta_a g k}{\omega} \cos(\omega t - kx + \phi) \frac{\cosh(k(z+h))}{\cosh(kh)},$$

$$u(t,x,z) = \frac{\zeta_a g k}{\omega} \sin(\omega t - kx + \phi) \frac{\cosh(k(z+h))}{\cosh(kh)}.$$

As stated before, linear theory can only be applied to very low waves. The range of suitability of linear theory in deep water is $H/\lambda < 0.0062$, with H the wave height and λ the wavelength. Linear wave theory can also be used to generate irregular seas, because the superposition principle can be applied. Given vectors of frequencies, amplitudes and phases, a linear wave can be built as the sum of the individual components.

Wave Kinematics Above the Calm Water Level

Since linear wave theory is only valid up to the calm water level and velocities are needed up to the free surface, some kind of stretching technique has to be used above the calm water level.

In this method Wheeler stretching has been used as a commonly accepted method, which is easy to implement. In the Wheeler stretching technique the negative z-axis has been extended from the actual instantaneous free surface elevation to the sea bed. This has been done by replacing z in the right-hand-side of Equations by

$$z' = \frac{h}{h+\zeta} z + h\left(\frac{h}{h+\zeta} - 1\right)$$

where

- z¢ is the computational vertical coordinate $-h \le z' \le 0$;
- z is the actual vertical coordinate $- \le z \le z$.

5th order Stokes Theory

For steeper waves, where in general the crests become higher and the troughs flatter, linear theory does no longer hold. To describe nonlinear waves, a solution to potential theory is used.

The solution is represented by Fourier series, and the coefficients in these series can be written as perturbation expansions with parameter Ak. Here, k is the wave number, which can be written in terms of the wave length $k - 2\pi/\lambda$, and A is the amplitude of the wave at lowest order.

The terms in the perturbation expansion can be found by satisfying boundary conditions on the free surface, and solving the resulting set of ordered equations.

The expansion of the series can be done, in theory, infinitely far, but in practice, the 5th order solution is already very complicated. Details about how to implement the 5th order Stokes solution can be found in, where the sign correction described by should be taken into account.

TREATMENT OF OPEN BOUNDARIES

Domain walls in wave simulations at open sea need to permit fluid flow in and out. Therefore, at least an inflow boundary is needed where the wave is generated, and an outflow boundary opposite to the inflow boundary where the wave leaves the domain. In the current method the waves, which are long crested, usually travel in the positive x-direction.

Because of the two-dimensional shape, the side boundaries can often be closed walls; when positioned far enough from the structure they do not influence the wave field in the close surroundings of the structure.

When the waves are highly distorted, such that side walls are severely influencing the flow, also outflow boundary conditions can be used at the side walls a cross-section of one row of cells is shown with the positions of the velocities at the inflow and outflow boundaries.

At the inflow boundary the velocities and VOF function in the column of cells directly left of the domain boundary are prescribed.

This way, the horizontal inflow velocity is positioned at the domain boundary, and no pressure is needed in the inflow cells. For the outflow boundary the horizontal velocity is shifted to one cell width away of the actual domain boundary.

Therefore, also a pressure is needed in the outflow cells, so velocities, pressure and VOF function need to be determined at the outflow boundary (using a Neumanntype outflow condition).

The wave descriptions have been given, from which velocities and wave height at the inflow boundary are derived. The remainder of this section deals with the outflow boundary.

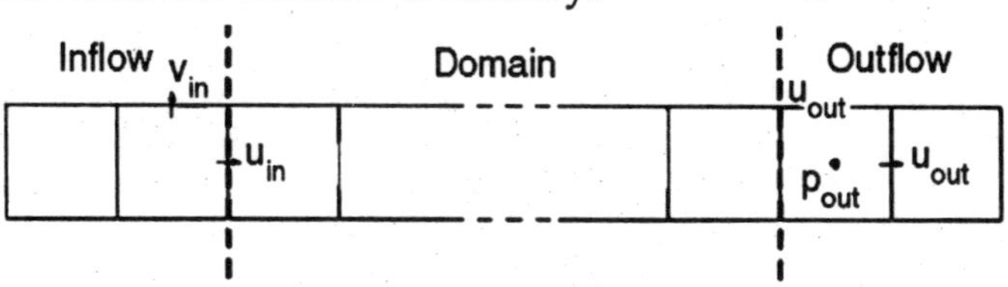

Fig. Position of Velocities at the Inflow Boundary, and Velocities and Pressure at the Outflow Boundary

Overview Outflow Boundary Conditions

A very important aspect of wave simulation is to determine the conditions at the outflow boundaries. If the wave is developing inside the domain, it should flow out of the domain as if there was no boundary. Otherwise, the wave will reflect from the boundaries into the domain, which disturbs the simulation. An overview of outflow boundary conditions that prevent wave reflections is given by Givoli.

The first class consists of special procedures for the numerical solution of wave problems in unbounded domains, that involve an artificial boundary but not the direct use of a non-reflecting boundary condition. Cerjan and others presented what can be termed a 'filtering scheme'. In this scheme, the amplitudes of the displacements are gradually reduced in a strip of nodes adjacent to the boundary.

Thus, the solution is artificially damped in the vicinity of the boundary. Another kind of dissipation zone was used by where an extra damping pressure was added to the free surface, which opposes the vertical wave velocity. A disadvantage of such damping zones is the increase of computational cells out of which the damping zone exists. Especially, in three dimensions many computational cells have to be added outside the real computational domain.

In the second class, which is the largest one, local non-reflecting boundary conditions (NRBCs) based on the scalar wave equation are used. There exist many variations in local NRBCs, every problem in different types of fields uses NRBCs that works best for that particular problem. A comparison of different boundary conditions derived from the discretisation of the multi-dimensional wave. The widely used NRBC of Sommerfeld is a discretisation of the one-dimensional scalar wave equation with an a priori chosen wave velocity.

A variation hereof was introduced by Orlanski, who calculated the wave velocity for every grid point and used that velocity in the discretised wave equation.

The advantage of this method is that no information is needed of the flow beforehand. Finally, non-reflecting boundary conditions are considered that are non-local in time or space or both.

These NRBCs have the disadvantage that many time levels should be stored in memory. The Sommerfeld and Orlanski boundary conditions is given. The implementation of a general non-reflecting boundary condition is described, and the damping zone used by is introduced.

Non-Reflecting Boundary Conditions

Boundary conditions based on the wave equation One way to prevent waves reflecting from the outflow boundary is to use a non-reflecting boundary condition. In case of waves most of the existing boundary conditions are based on the wave equation

$$\frac{\partial \phi}{\partial t} + c\frac{\partial \phi}{\partial x} = 0,$$

where f is any quantity that travels wave-like as velocity or pressure, and c the wave velocity. In Equation the direction of the wave is in the x-direction. In the Sommerfeld boundary condition the wave velocity is chosen a priori, which is easy when regular waves are simulated.

In that case, the Sommerfeld condition gives very good results as will be shown later. When irregular waves are considered, or when the regular waves are deformed due to the presence of an object, the method does not give very accurate results, since only one wave velocity can be chosen.

For those situations, Orlanski developed a method where the wave velocity is not chosen a priori, but is calculated every time step based on the local wave kinematics near the outflow boundary. The problem with this method is to determine the wave velocity accurately.

To calculate c, first the wave equation is discretised using finite differences in cells close to the outflow boundary where the condition will be imposed. The wave equation is discretised for the horizontal velocity in every level of cells in the water depth. The c is then calculated as the mean of the approximated wave velocities in every level in the water depth. The calculated wave velocity is shown for a regular wave with an actual wave velocity of 22.6 m/s during one wave period.

The calculated wave velocity is not constant at all, but oscillates around the theoretical value. The jump that occurs around 7.2 s is present because there the crest of the wave travels through the outflow boundary resulting in ¶*u*/¶ = 0. This can cause the wave velocity to jump from minus infinity to plus infinity.

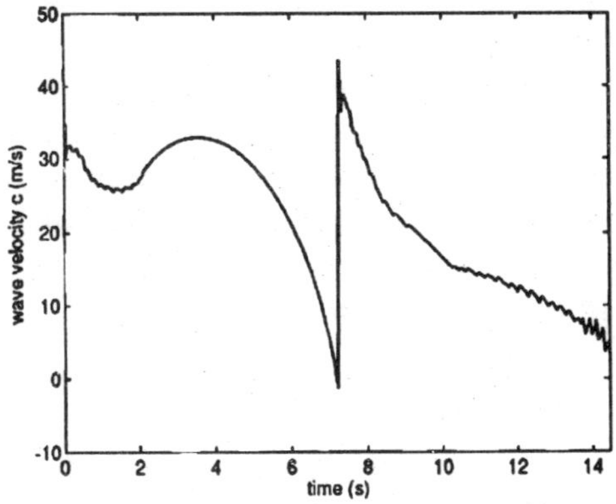

Fig. Calculated Wave Velocity Using a Finite Difference Approximation of the Wave Equation

In this region, the wave velocity is adapted, such that these extreme values are not used in the outflow boundary conditions. In an investigation of wave propagation using Orlanski's method at the outflow boundary, it turned out that the results were not very accurate, so this method is not used in the simulations shown in this thesis.

The Implementation of a Non-Reflecting Boundary Condition

At the outflow boundary conditions for pressure and velocity are needed. When using a Sommerfeld boundary condition, for the velocities at the outflow boundary, the wave equation Equation is discretised using finite differences in the following way

$$\frac{\phi_e^{n+1} - \phi_e^n}{\delta t} + C\frac{\phi_e^n - \phi_c^n}{\delta x_e} = 0,$$

and ϕ is used for the velocity components. From this equation the velocities at the outflow boundary u_e^{n+1}, v_e^{n+1} and w_e^{n+1} follow.

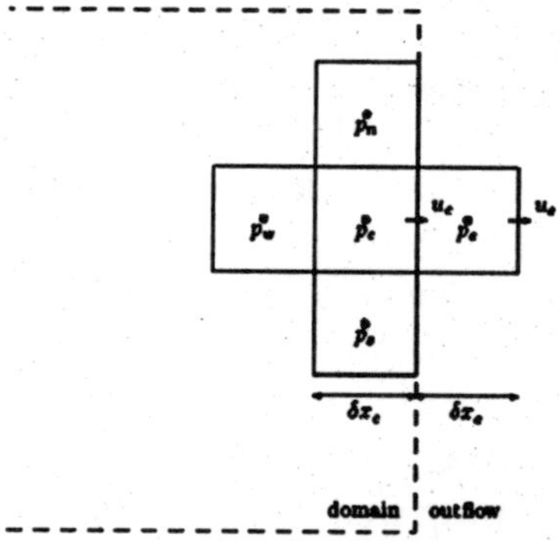

Fig. Configuration of Cells Near the Outflow Boundary

For the calculation of the pressure in the outflow cell the procedure is somewhat more complicated, since the pressure appears in the Poisson equation that is solved using an iterative method.

To avoid an iteration in the outflow cell, the boundary condition is substituted in the equation for the interior pressure cell pe, such that the equation for the pressure in the outflow cell is not needed.

At the end of the iterations the pressure in the outflow cell is updated using the pressure in the interior cell. This procedure is elaborated for a general outflow boundary condition.

$$\alpha\frac{p_e - p_c}{\delta x_{pe}} + \beta p_e = cA + \beta p_0,$$

Here, the Neumann condition can recognised for (α, β) = (1, 0) and the Dirichlet condition for (α, β) = (0, 1). The Sommerfeld condition is obtained when taking $\alpha = c$, $\beta = 1/\delta t$, $A = 0$ and $p_0 = p_e^n$.

This equation is discretised using finite differences

$$\alpha \frac{p_e - p_c}{\delta x_{pe}} + \beta p_e = \alpha A + \beta p_0,$$

with $\delta x_{pe} = (\delta x_e + \delta x_c)/2$. From this equation the pressure in the outflow cell pe can be solved

$$p_e = \frac{\alpha}{\alpha + \beta \delta x_{pe}} p_c + \frac{\alpha A \delta x_{pe} + \beta p_0 \delta x_{pe}}{\alpha + \beta \delta x_{pe}}.$$

$$C_c p_c + C_w p_w + C_e p_e + C_n p_n + C_s p_s = RHS.$$

The coefficients Cc to Cs contain squared grid sizes and the right-hand-side RHS contains the divergence of the contributions from convection, diffusion and external forces. In this equation the outflow pressure given by Equation can be substituted resulting in

$$\left(C_c \frac{\alpha}{\alpha + \beta \delta x_{pe}} C_e \right) p_c + C_w p_w + C_n p_n + C_s p_s$$

$$= RHS - \frac{\alpha A \delta x_{pe} + \beta p_0 \delta x_{pe}}{\alpha + \beta \delta x_{pe}} C_e.$$

The matrix containing the coefficients remains diagonal dominant when α, $\beta > 0$. After the pressure in the interior of the domain is solved from the Poisson equation, the pressure in the outflow cell is updated using Equation.

Pressure Damping at the Free Surface

Instead of using a non-reflecting boundary condition, a dissipation zone can be used where the wave is damped. In the current method, for the damping inside the dissipation zone a pressure term has been added to the free surface pressure. Physically, this can be interpreted as the air acting like a damper on the wave. The pressure added to the atmospheric pressure term at the free surface is chosen as a function of the vertical velocity at the free surface:

$$pdamp(t, x, \zeta) = \alpha(x)\, w(t, x, \zeta).$$

The damping function $\alpha(x)$ should be chosen such that the wave is damped completely, and the wave should not reflect at the start of the damping zone. A polynomial form for the function is used by concludes in his report that a linear function is suitable for the purposes of wave simulation. If a linear damping function $\alpha(x) = ax + b$ is used, two constants a and b have to be chosen. Here, a is the slope of the damping function, which determines the rate of the damping. The constant b has to be chosen such that the damping function is zero at the start of the dissipation zone.

The slope of the damping function is determined by the characteristics of the wave that is to be damped. Meskers shows in his report how the slope and the length of the dissipation zone can be determined after choosing the total reflection that is allowed. For a number of allowed reflection factors the slope and length of the dissipation zone are plotted.

For regular waves, the following steps have to be taken to determine the slope and length of the dissipation zone. First, choose the amount of reflection (rtot) that can be permitted. This amount of reflection is calculated theoretically, and is only valid for a perfectly regular wave.

Then, given the wave frequency and reflection coefficient, determine the length of the numerical beach using the curved lines. At the left coordinate axis, the accompanying values are placed. Finally, determine the slope of the numerical beach, using the straight lines with the values of the right coordinate axis.

Besides the choice of the function, there also are some different possibilities for the closing wall of the domain, by which we mean the wall opposite of the inflow boundary.

- In the first option the wave is damped completely, the closing wall of the domain is a solid wall, no water is flowing out.
- In the second option the wave is damped towards its analytical form. So a 'perfect' wave is formed at the end of the domain. The closing wall is an outflow boundary, at which the analytical velocities are prescribed. This works very nicely in the case of wave propagation simulations without an object that disturbs the wave.

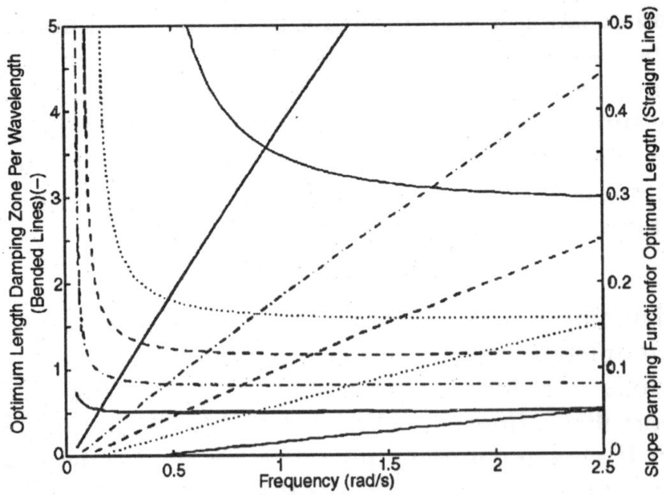

Fig. Combination of Required Length and Slope of the Dissipation Zone for a Certain Allowed Amount of Reflection.

A combination of the dissipation zone with a Sommerfeld boundary condition is used. In this case, the closing wall is an outflow boundary where a Sommerfeld boundary condition has been applied. The Sommerfeld boundary condition is tuned for the smallest frequencies of the wave in the domain, whereas the dissipation function is tuned for the larger frequencies.

In the current model we aim for the most general solution, so damping towards zero (the first option) is used. The wave is only damped in the travelling direction, whereas the tangential sides of the domain are solid walls, which have been placed at such a distance that the simulation results are not influenced by them.

When applying this to a rather high wave, some problems arise. The water level in the domain increases linearly in time, which is due to the fact that the net amount of water flowing into the domain at the inflow boundary is positive over one period and not zero, the so-called Stokes drift where the results of a wave simulation are shown with a period of 14.44 s, a wave height of 32.6 m, a wavelength of 325 m and the water depth is 600 m.

In the left of the figure the amount of water flowing into the domain is shown. Clearly, the integral over one period is not zero but positive. Because there is a solid wall at the end of the domain, the total amount of water is determined by the in- and outlet at the inflow boundary only, which causes the water level to increase linearly in time as is shown in the right of the figure. Because the wave is quite high, there is a very large increase of the water level, which is directly reflected in the mean of the wave elevation.

There are also some problems near the inflow boundary, where the wave is disturbed. Due to the reflections from the outflow boundary and the rise of the water level, the velocities and water height in the interior of the computational domain do not fit to the analytically prescribed velocities at the inflow boundary any more.

There are several ways to solve this increase of water problem. One way is to determine the net inflow and add an outflow boundary where this amount of fluid is forced out of the domain. A disadvantage of this method is, that it can only be applied to regular waves in a nice way, in which case it is known how much fluid should flow out of the domain at what time, due to the regularity.

A more flexible method has been found by changing the solid wall at the end of the domain into an outflow boundary, where hydrostatic pressure is prescribed. In a perfect situation, the water height at the end of the domain always equals the calm water level because of the damping of the wave. The fluid simulation will always try to maintain that level when the hydrostatic pressure for a water level equal to the water depth is prescribed at the outflow boundary.

Instead of the linear increase of water, a fluctuation of the water volume around the initial amount of water can be observed. The fluctuation is due to some reflections, which can be concluded from the fact that the period of the fluctuation is 8 wave periods, which means that the reflected wave travels 8 wavelengths before it reaches the starting position, the outflow boundary, again. In a domain consisting of 4 wavelengths, this is exactly 8 wavelengths (back and forth in the domain). The increase and decrease of the total water level during 8 periods is less than 0.5%.

FREE SURFACE VELOCITIES

The treatment of the velocities in the neighbourhood of the free surface is explained. It was concluded that special care has to be taken for the determination of SE-velocities, which are velocities at the cell face between a surface cell and an empty cell. Two methods are described, of which a combination is used in COMFLOW. In the first method SE-velocities are defined by demanding conservation of mass in an S-cell. This means that the total flux through the cell faces of the S-cell should be zero.

Another disadvantage of this method is the inaccuracy in wave simulations. The inaccurate prediction of an SE-velocity in case of a wave simulation can be understood from the right of the figure the horizontal velocity has been shown as function of the vertical coordinate. The theoretical values of the horizontal velocity in the neighbourhood of the free surface are indicated by the solid line.

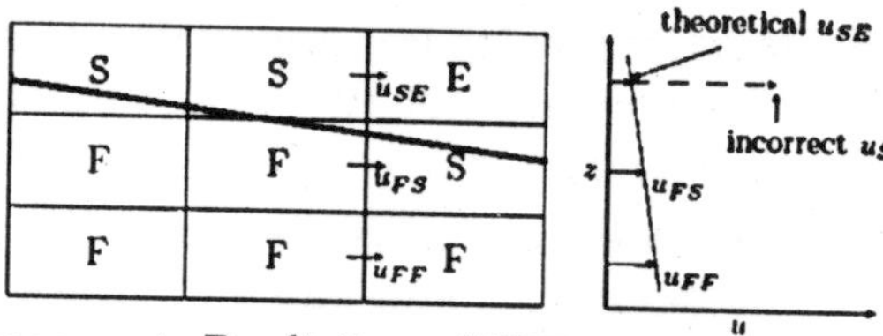

Fig. Inaccurate Prediction of USE in a Wave Simulation

To satisfy div(u) = 0 in the central S-cell, the SE-velocity uSE is copied from the left neighbour velocity (and the vertical velocity of this S-cell vSE is copied from the lower cell face). Due to the coarseness of the grid, this neighbour velocity is about one or more meters left of the SE-velocity resulting in an inaccurate prediction of u_{SE} sketched by the dashed arrow.

In the second method for SE-velocities it is proposed to determine the velocities by choosing a direction, from which all SE-velocities in an

S-cell are extrapolated. This direction is chosen as the direction where most fluid is present, so the coordinate direction that is 'most normal' to the free surface.

In the case of waves, this is mostly the negative z-direction. As stated before, constant and linear extrapolation can be used. In wave simulations linear extrapolation gives superior results, since then the velocities are estimated very accurately. Here, an irregular steep wave propagates through an empty domain. A snapshot of the wave elevation is shown at the time point that the wave is most steep.

The asterisks show measurements of this wave. The mass conservation method does not predict the wave elevation accurately for this steep wave. Both constant and linear extrapolation produce a much better resemblance with the measurements. The simulation using linear extrapolation is most accurate. Based on the simulation of waves.

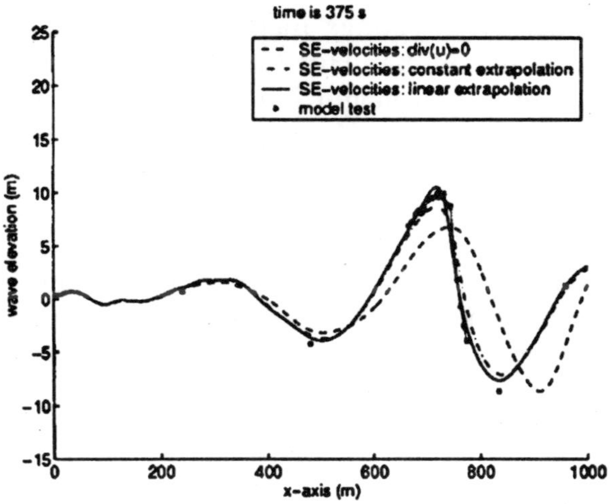

Fig. Wave Elevation of an Irregular Wave where Different SE-Velocity Treatments have been Compared

WAVE PROPAGATION USING DIFFERENT OF METHODS

A few different VOF methods have been described and their performance has been tested using standard kinematic tests and a dambreak simulation. Four different methods have been used. First, the original Hirt-Nichols method has been used, where the interface is implicitly reconstructed in a piecewise constant manner.

To avoid flotsam and jetsam and to take care of mass conservation, a local height function has been introduced. A more sophisticated method for the displacement of the free surface is the method of Youngs. The free surface is reconstructed using piecewise linear elements and the displacement is based on this reconstruction. Youngs' method has also been used in combination with the local height function to ensure perfect mass conservation. The extrapolation method for free surface velocities has been used.

First, the different VOF methods are tested on the propagation of a regular wave. The wave has a period of 14.44 seconds and a length of 325 meter. The wave height is 10.14 meter and the water has a depth of 600 meter. At the outflow boundary a Sommerfeld condition is used. The resulting free surface profile after a simulation time of four wave periods. Also the error in the calculated free surface profile compared to the theoretical solution is shown, calculated by

$$E_w(x) = |\,\eta(x) - \eta th(x)\,| / H,$$

where η is the calculated wave elevation, η*th* the theoretical wave elevation, and H the wave height. From the resulting error Ew, plotted in the bottom concluded that Youngs' method gives more accurate results in a regular wave simulation than Hirt-Nichols' method.

Youngs without a local height function is best in the interior of the domain, but has a larger error at the outflow boundary than Youngs with a local height function. No reason has been found for this behaviour. In all four simulations relatively little mass was lost. When examining the total amount of water in an area of 30 m high around the calm water level, all four methods have mass loss within 0.5% of the amount of water in that area. The calculation times of the four simulations are comparable. The factor between the calculation times of Youngs and Hirt-Nichols is 1.2.

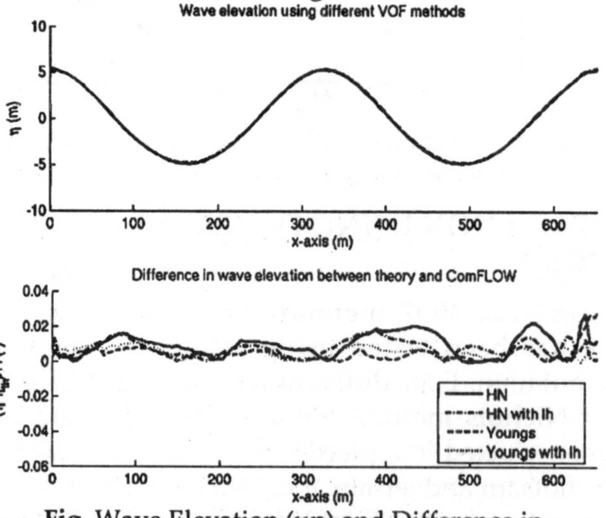

Fig. Wave Elevation (up) and Difference in Wave Elevation between Stokes 5th Order Theory and Simulation (down) for Different VOF Methods

Youngs method does not take much extra calculation time in this simulation, since the interface reconstruction is only performed in a small percentage of the total number of cells (in average in 130 of the 6000 cells a reconstruction is made). The four different VOF methods are also used in the propagation of a steep wave event. The wave event is taken from an experiment at MARIN, to be more precise experiment 114002.

The wave is generated at the inflow boundary by prescribing a superposition of linear wave components derived from Fourier analysis of the measured wave elevation in the experiment. The wave elevation at the position and time point where the wave is high and steep. Youngs' method without a local height function results in the highest wave and gives the best agreement with the measurement.

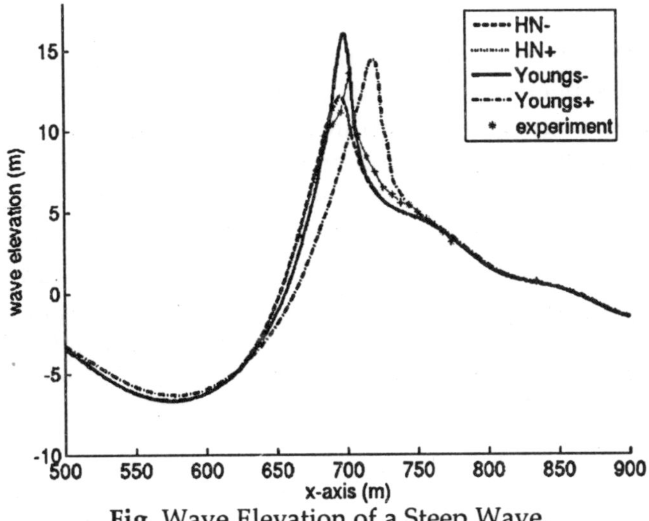

Fig. Wave Elevation of a Steep Wave.

Hirt-Nichols' VOF and Youngs' VOF with and without a local height function Hirt-Nichols with and without local height function give similar results, both are pretty good, but the wave crest is a bit flattened compared to the experiment. Youngs with a local height function performs worst, which is remarkable considering the good performance.

To conclude, Youngs' method performs best in wave simulations, its results are most accurate. A problem arises when combining Youngs' method with a local height function in the steep wave event. So, this combination should only be used with care until this problem is solved. Unfortunately, due to the choice of free surface velocities that are extrapolated, Youngs' method without local height function can cause loss of mass in the dambreak.

But in a simulation with a very smooth free surface as in the wave simulations, only a very small amount of water is lost using Youngs' method. So, for wave simulations, Youngs' method without local height function can be used without problem. But when performing a more violent simulation with a much distorted free surface, mass can be lost just mentioned.

Mass can be perfectly conserved using Youngs' method, when it is combined with div(u) = 0 in surface cells. But using div(u) = 0 for the determination of free surface velocities results in inaccurate wave simulations and can cause instabilities in the computations. Therefore, this option should not be used. The safest way is to use Hirt-Nichols' method with local height function, but this can be a bit less accurate.

VALIDATION OF WAVE PROPAGATION

For the validation of wave propagation in several tests have been performed. Firstly, two-dimensional wave propagation without an object in the flow has been investigated. Attention has been paid to reflections at the outflow boundary, the influence of the artificial viscosity present due to the upwind discretisation and the size of the grid and time step necessary for an accurate simulation of waves.

Especially, steep and high waves, which are the most important waves in green water and wave impact calculations, have been studied. Secondly, simulations have been performed of wave loading on a spar platform. The waves are regular, very long and quite low. The simulation results have been compared with experimental results that have been provided by the Maritime Research Institute Netherlands (MARIN).

Two-Dimensional Wave Propagation: Regular Waves

A very extensive study of two-dimensional wave propagation without an object in the flow has been performed by Meskers. Some of the most important results will be shown here also. Attention will be paid to the number of cells and the time step that are needed for an accurate description of the wave.

The simulations have been performed and compared with both the linear and the 5th order Stokes theory. Also, the influence of the artificial viscosity has been studied by performing a simulation of many wavelengths during many periods. These waves have been studied by simulating a design wave, which is a kind of mean shape of a wave in a linear random sea-state of which the power spectrum is given. Also a comparison of the use of a Sommerfeld boundary condition with a damping zone is made.

The simulations have been performed for one example of a wave, which is one of the characteristic waves in an FPSO field with a water depth of 600 meter. This wave has a period of 14.44 seconds, resulting in a wave length of 325 meter.

The wave height is varied between 10.14 and 20.28 meter. From the conclusions of Meskers in the characteristic parameter settings that have shown to give an accurate simulation of this example wave can also be used for deep water waves with different periods and wavelengths.

In the simulations of the wave shown in this thesis, a few numerical parameters have been varied, namely the number of time steps per period, the number of cells per wavelength, and the number of cells in the wave height. The results have been presented as the resulting wave elevation after a simulation time of four periods.

The domain consists of four wavelengths in case of using a damping zone, of which two wavelengths are used as damping zone. The domain consists of two wavelengths when the Sommerfeld boundary condition is used.

Airy wave theory and 5th order Stokes theory have been used for the initial condition and the inflow boundary condition. In the lower picture the difference between the simulation and linear theory and Stokes 5th order theory, respectively, has been shown.

Clearly, the 5th order Stokes results are much better than the linear Airy wave results, which is consistent with the fact that H/_ equals 0.06, outside the validity region of linear theory. First, the Sommerfeld outflow boundary condition is used that is based on the wave equation. Second, a damping zone is added as in the other simulations. Both methods give similar results.

The Sommerfeld boundary condition lets the wave flow out of the domain properly without much disturbance in the domain. The advantage of using the Sommerfeld condition over a damping zone is the number of grid cells that need to be used.

In this simulation, the number of grid cells in the simulation using a damping zone is twice as large as the number of grid cells in the Sommerfeld simulation.

On the other hand, the Sommerfeld condition can only be used in case of regular waves that are not too much disturbed. Further, the wave velocity should be given a priori, which is only possible when the wave characteristics are known.

The number of time steps per period has been altered. The difference between 250 and 500 time steps per period is not very large. From this it is concluded that 250 time steps per period is enough for an accurate simulation result.

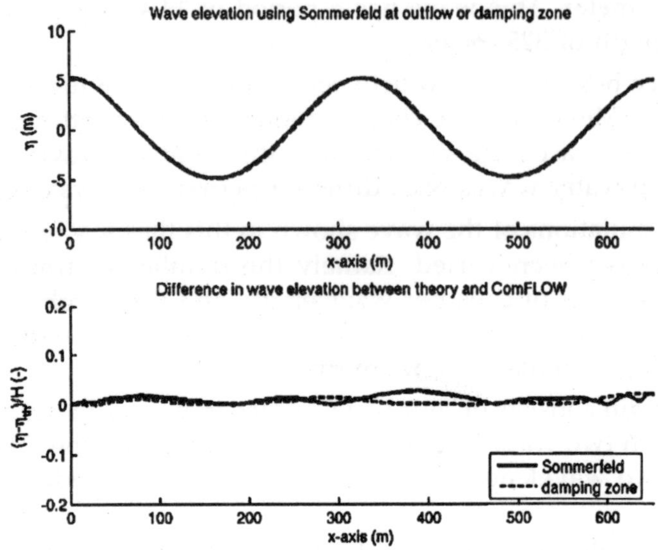

Fig. Wave Elevation (Top) and Difference in Wave Elevation

The accuracy of the wave simulation increases with a larger number of cells per wavelength. There is no difference between 60 and 90 cells per wavelength, so 60 cells per wavelength is enough to capture the wave of altering the number of cells in the wave height. The grid was equally stretched towards the calm water surface in the three different simulations.

There is not a large difference between the results. For all three simulations, the difference between computation and theory is about 5% of the wave height. In the case of 9 cells in the wave height, the simulation shows some small peaks, especially in the crest and trough of the wave.

This is due to the ratio between dx and dz, which are the distances between two grid lines in x and z direction, respectively. Meskers gives in his report an estimate for the minimum required aspect ratio to prevent these wiggles, based on a series of simulations. In the future, this point should be investigated further to understand the nature of the wiggles. Dissipation of energy in wave simulations. Compared to a central discretisation the upwind discretisation can be interpreted as a central discretisation plus an extra diffusive term. This term adds extra viscosity to the physical viscosity.

The amount of extra viscosity is equal to uh/2 where u is the velocity and h the mesh size. So this term is dependent on the position in space. When simulating waves, we have to investigate the influence of the artificial diffusion on the wave propagation. It can be seen that the influence is definitely not very large over a few periods, because no real damping is visible after the four periods that have been simulated. To get a better idea of the influence of the artificial viscosity, a wave has been simulated for many periods in a domain of 25 wavelengths.

The wave has a period of 1.9 seconds and a wave height of 0.16 meter. The wave elevation as function of time and as function of distance to the inflow boundary are shown. The wave elevation as function of time at a position 14 wavelengths from the inflow boundary is shown. The simulation is started with an undisturbed (and thus undamped) wave field.

During the simulation the amplitude of the wave decreases, until the stationary damping is reached after about 58 seconds. At this position, 14 wavelengths from the inflow boundary, the wave height has decreased from 0.16 meter to 0.125 meter, which is a decrease of 22% of the wave height.

From the right picture it can be seen that the wave height has decreased 28% at 20 wavelengths behind the inflow boundary after a simulation time of 100 periods. An estimate of the maximum artificial viscosity that is added due to the upwind discretisation can be found by calculating uh/2.

For the velocity u the maximum of the mean velocity over one period is taken, which according to linear theory is equal to

$u = gAk/\omega \approx 0.26$m/s.

So the approximate maximum artificial viscosity is given by

$k_{art} = uh/2 \approx 0.26 * 140/1500/2 = 0.012$ n.

Where as the physical kinematic viscosity v is equal to 10^{-6} m^2/s is really caused by the artificial viscosity, simulations have been performed with less artificial viscosity (only 10% of the artificial viscosity is added) and with a central discretisation.

The simulation with the central discretisation has been performed for 58 periods in a domain of 12 wavelengths, which is because the grid sizes and time step have to be very small in a calculation with a central discretisation.

The wave elevation at 9 wavelengths behind the inflow boundary has been given for the different amounts of added viscosity. It can be clearly seen, that the amplitude of the wave is decreasing more in time when the added viscosity becomes larger where the wave height decrease is shown after 100 periods of simulation at the different locations behind the inflow boundary.

We can conclude from this figure that there is a large influence of the artificial viscosity, which has a damping effect on the wave.

But clearly, this is not the only reason for dissipation of energy in the wave simulations. Although in the central discretisation no artificial viscosity is present, the wave has been damped much more than it should be according to the physics.

This is due to the boundary conditions and the displacement of the free surface that sometimes are chosen to be a bit dissipative to get a stable solution. Summarising, for wave simulation in a very long domain, where many periods are simulated, a clear damping is visible.

This dissipation of energy is for a great deal due to the artificial viscosity that is added when using an upwind discretisation. But another part of the energy dissipation is coming from the treatment of the boundaries and the free surface.

Although an energy preserving discretisation is used, some energy is lost in other parts of the algorithm. The loss of energy is only a few per cent in a small domain when not that many periods are simulated. So for the applications of wave loading on ships, the influence will not be very significant.

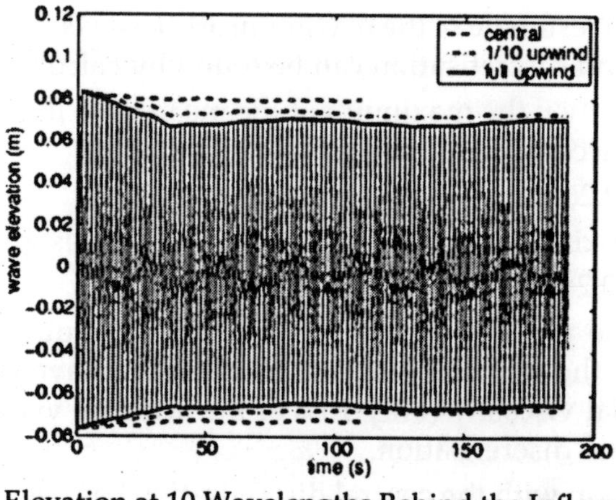

Fig. Wave Elevation at 10 Wavelengths Behind the Inflow Boundary

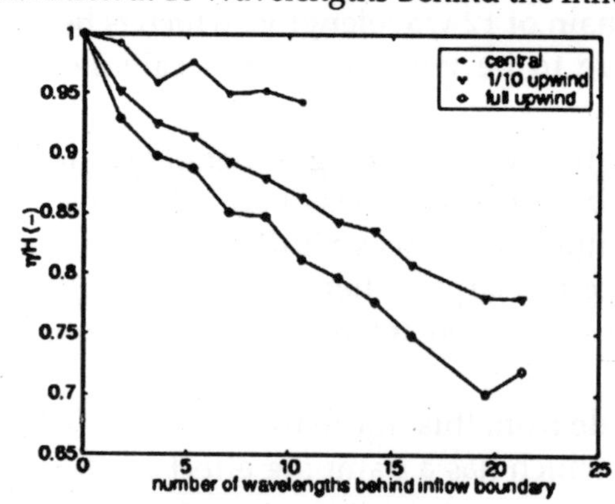

Fig. Decrease of Wave Height after A Simulation Time of 100 Periods at Different Distances from the Inflow Boundary

Two-Dimensional Wave Propagation: Steep Wave Events

For the validation of irregular steep or high waves, waves generated on basis of Newwave theory have been used such a wave event has been developed as a mean wave shape of a high wave in a given linear random sea-state. The wave is modelled by a superposition of linear waves, and thus is a linear wave. High waves are not linear of form, and therefore, mostly some nonlinear corrections are being made to the waves.

The waves are sometimes used as a design wave, because they have the nice feature that the position and time of the largest impact can be predicted beforehand. At marin these design waves have been used in the experimental programme of the Safe flow project to get a better understanding of wave slamming.

Before the experiments with a vessel in the waves is performed, the waves have been calibrated in the basin without the presence of a vessel. The measurements of this undisturbed wave elevation have been used to compare our numerical method.

Table. Characteristics of the Wave Events, for which the Simulations are Compared with Measurements

Test no.	Single wave event description
112003	1/14 stepp wave in sea state steepness for 100 year return period
113001	mean highest wave for 100 year return period
114002	1/16 steep wave in sea state steepness for 100 year return period
119001	1/18 steep wave in sea state steepness for 100 year return period
120001	highest wave in design wave spectrum with Hm_0 = 12m, T_0 = 12s, γ = 2.5.
121001	steepest wave in design wave spectrum with Hm_0 = 12m, T_0 = 12s, γ = 2.5.

The waves have been generated at the inflow boundary of the numerical wave tank by using a Fourier transform of the time trace of a measured wave height. The calculated wave elevation in the domain is compared to the measurements at positions of 240, 480, 720, and 960 m behind the inflow boundary.

Detailed measurements, using 19 wave probes, have been performed at the position where the wave is expected to be highest between 666 and 774 m behind the inflow boundary. The calculated wave height of test 112003 at 720 m behind the inflow boundary, where the wave is expected to be highest, is compared to theory and experiment. The comparison with the experiment is very good. The amplitude of the simulation is a bit smaller than in the experiment, which has been observed before as a consequence of using Hirt-Nichols' method for the free surface displacement.

Also the dissipation of energy due to the upwind discretisation could have an influence on the wave height. The simulation has been performed on two different grids, using 600 × 50 grid cells and 1200 × 100 grid cells.

The grid refinement does not have a large influence, almost no difference can be observed between the two grids. The wave elevation predicted from linear theory, extrapolated from the FFT of the measured wave elevation, is shown, which is very different from the calculations and measurement.

This means that Com FLOW has correctly dealt with the nonlinearities that are present in reality. The simulations of the other events have all been performed with the fine grid of 1200 × 100 grid points are much the same as the results of the first test. The wave does not completely reach the height of the experiment in the first peak.

This is confirmed by the wave profile at 430 s, where the wave has collapsed a little. Again, the simulations do not reproduce the exact height of the first peak at about 395 seconds in both tests. Nevertheless, the other parts of the simulation are fairly close to the experiment.

To be sure that the wave is well reproduced in the simulation the wave must be compared with the undisturbed wave in the basin first. When it has been confirmed that the wave is the same as in the basin, it can be used for a simulation of wave impact.

Further, the results show that Com FLOW reproduces nonlinearities from the experiment, but does not always reproduce the height of the wave exactly.

Regular Wave Loading on a Spar Platform

For the validation of wave loading, simulations have been performed of regular wave loading on a spar platform. A spar platform is a floating structure for drilling and the production of crude oil in the ocean. Typically, a spar is a long cylindrical steel structure with 30-50 meters in diameter and 200 meters in length.

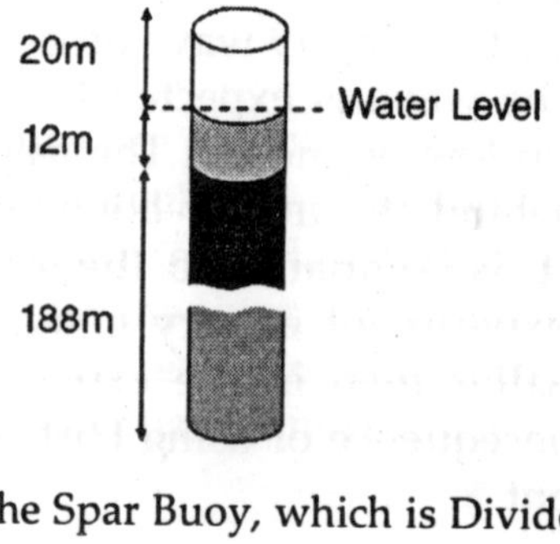

Fig. Configuration of the Spar Buoy, which is Divided into 3 S®segments.

At MARIN experiments with a spar have been performed, where the spar was fixed while regular waves hit the structure. In full scale the spar has a total length of 220 m and a diameter of 35 m, the draft is 200 m in a water depth of 290.35 m. The spar has been divided into three horizontal segments, on which forces have been measured.

The spar buoy has been placed at a distance of one wavelength behind the inflow boundary.

Table. Wave Characteristics of Simulations of Loading on a Spar Platform.

Test No.	Frequency	Period	Wave height	Wavelength
101	0.26 rad/s	24.17s	10.77 m	866 m
201	0.26 rad/s	24.17s	23.62 m	866 m
001	0.48 rad/s	13.09s	10.91 m	274 m

About 100 m behind the cylinder, a dissipation zone of one wavelength has been added to prevent the wave from reflecting into the domain.

Chapter 10

Lasers

"Laser" is an acronym for "Light Amplification by Stimulated Emission of Radiation", coined in 1957 by the laser pioneer Gordon Gould. Although this original meaning denotes an principle of operation, the term is now mostly used for *devices* generating light based on the laser principle.

The first laser device was a pulsed ruby laser, demonstrated by Theodore Maiman in 1960. In the same year, the first gas laser (a helium–neon laser) and the first laser diode were made. The term "optical maser" (MASER = microwave amplification by stimulated amplification of radiation) was initially used, but later replaced with "laser".

Laser technology is at the core of the wider area of photonics, essentially because laser light has a number of very special properties:

- It is usually emitted as a laser beam which can propagate over long lengths without much divergence and can be focused to very small spots.
- It can have a very narrow bandwidth, whereas e.g. most lamps emit light with a very broad spectrum.
- It may be emitted continuously, or alternatively in the form of short or ultrashort pulses, with durations from microseconds down to a few femtoseconds.

These properties, which make laser light very interesting for a range of applications, are to a large extent the consequences of the very high degree of coherence of laser radiation.

LASER WORKS

A laser usually comprises an optical resonator (laser resonator, laser cavity) in which light can circulate (e.g. between two mirrors), and within this resonator a gain medium (e.g. a laser crystal), which serves to amplify the light. Without the gain medium, the circulating light would become weaker and weaker in each resonator round trip, because it experiences some losses, e.g. upon reflection at mirrors.

However, the gain medium can amplify the circulating light, thus compensating the losses if the gain is high enough. The gain medium requires some external supply of energy – it needs to be "pumped", e.g. by injecting light (*optical pumping*) or an electric current (*electrical pumping* '! *semiconductor lasers*). The principle of laser amplification is stimulated emission.

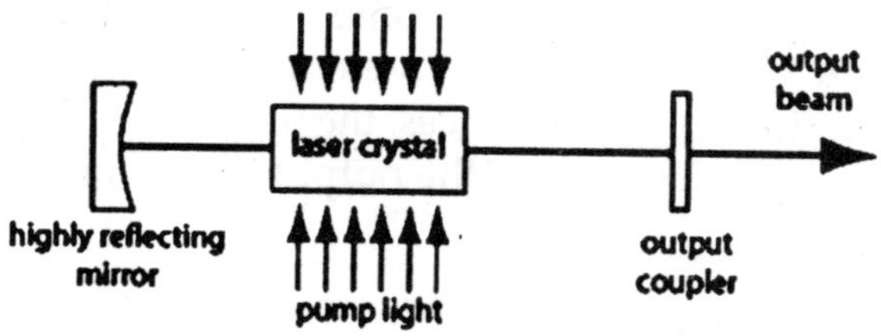

Fig Setup of a Simple Optically Pumped Laser.

The laser resonator is made of a highly reflecting curved mirror and a partially transmissive flat mirror, the output coupler, which extracts some of the circulating laser light as the useful output. The gain medium is a laser crystal, which is side-pumped, e.g. with light from a flash lamp.

A laser can not operate if the gain is smaller than the resonator losses; the device is then below the so-called laser threshold and only emits some luminescence light. Significant power output is achieved only for pump powers above the laser threshold, where the gain can exceed the resonator losses.

If the gain is larger than the losses, the power of the light in the laser resonator quickly rises, starting e.g. with low levels of light from fluorescence. As high laser powers saturate the gain, the laser power will in the steady state reach a level so that the saturated gain just equals the resonator losses (' *gain clamping*). Before reaching this steady state, a laser usually undergoes some relaxation oscillations. The threshold pump power is the pump power where the small-signal gain is just sufficient for lasing.

Some fraction of the light power circulating in the resonator is usually transmitted by a partially transparent mirror, the so-called output coupler mirror. The resulting beam constitutes the useful output of the laser. The transmission of the output coupler mirror can be optimized for maximum output power.

Some lasers are operated in a continuous fashion, whereas others generate pulses, which can be particularly intense. There are various methods for pulse generation with lasers, allowing the generation of pulses with durations of microseconds, nanoseconds, picoseconds, or even down a few femtoseconds (→ *ultrashort pulses* from *mode-locked lasers*).

The optical bandwidth (or linewidth) of a continuously operating laser may be very small when only a single resonator mode can oscillate ('! *single-*

frequency operation). In other cases, particularly for mode-locked lasers, the bandwidth can be very large – in extreme cases, it can span about a full octave.

The centre frequency of the laser radiation is typically near the frequency of maximum gain, but if the resonator losses are made frequency-dependent, the laser wavelength can be tuned within the range where sufficient gain is available. Some broadband gain media such as Ti:sapphire and Cr:ZnSe allow wavelength tuning over hundreds of nanometers. Due to various influences, the output of lasers always contains some noise in properties such as the output power or optical phase.

Types of Lasers

Common types of lasers are:

- *Semiconductor lasers* (mostly laser diodes), electrically (or sometimes optically) pumped, efficiently generating very high output powers (but typically with poor beam quality), or low powers with good spatial properties (e.g. for application in CD and DVD players), or pulses (e.g. for telecom applications) with very high pulse repetition rates. Special types include quantum cascade lasers (for mid-infrared light) and surface-emitting semiconductor lasers (VCSELs and VECSELs), the latter also being suitable for pulse generation with high powers.
- *Solid-state lasers* based on ion-doped crystals or glasses (*doped insulator lasers*), pumped with discharge lamps or laser diodes, generating high output powers, or lower powers with very high beam quality, spectral purity and/or stability (e.g. for measurement purposes), or ultrashort pulses with picosecond or femtosecond durations. Common gain media are Nd:YAG, Nd:YVO_4, Nd:YLF, Nd:glass, Yb:YAG, Yb:glass, Ti:sapphire, Cr:YAG and Cr:LiSAF. A special type of ion-doped glass lasers are:
- *Fibre lasers*, based on optical glass fibers which are doped with some laser-active ions in the fibre core. Fibre lasers can achieve extremely high output powers (up to kilowatts) with high beam quality, allow for widely wavelength-tunable operation, narrow linewidth operation, etc.
- *Gas lasers* (e.g. helium–neon lasers, CO_2 lasers, and argon ion lasers) and excimer lasers, based on gases which are typically excited with electrical discharges. Frequently used gases include CO_2, argon, krypton, and gas mixtures such as helium–neon. Common excimers are ArF, KrF, XeF, and F_2.

Less common are chemical and nuclear pumped lasers, free electron lasers, and X-ray lasers.

Laser Sources in a Wider Sense

There are some light sources which are not strictly lasers, but are nevertheless often called *laser sources*:

- In some cases, the term is used for amplifying devices emitting light without an input (excluding seeded amplifiers). An example is X-ray lasers, which are usually *superluminescent sources*, based on spontaneous emission followed by single-pass amplification. There is then no laser resonator. A similar situation occurs for optical parametric generators, where the amplification, however, is not based on stimulated emission. Light from such devices can have laser-like properties, such as strongly directional emission and a limited optical bandwidth.
- In other cases, the term *laser sources* is justified by the fact that the source contains a laser, among other components.
  This is the case for combinations of lasers and amplifiers ('! *master oscillator power amplifier*), and also for sources based on nonlinear frequency conversion of laser radiation, e.g. with frequency doublers or optical parametric oscillators.

SAFETY ASPECTS

The work with lasers can raise significant safety issues. Some of those are directly related to the laser light, in particular to the high optical intensities achievable, but there are also other hazards related to laser y for details.

Laser Light

Laser light (laser radiation) is simply light generated with a laser device. Such light has some very special properties, which very much distinguish it from light with other origins:

- Laser light is usually delivered in the form of a laser beam, i.e. it propagates dominantly in a well-defined direction with moderate beam divergence. Such a laser beam has a high (sometimes extremely high) degree of spatial coherence.
  This means that the electric fields at different locations across a beam profile oscillate with a rigid phase relationship. Exactly this coherence is the reason why a laser beam can propagate over long distances without spreading very much in the transverse directions, and why it can be focused to very small spots (high *focusability* of laser beams).
- In many but not all cases, laser light also has a high degree of *temporal* coherence, which is equivalent to a long coherence length. This means that a rigid phase relationship is also maintained over

relatively long time intervals, corresponding to large propagation distances (often many kilometers) or to huge numbers of oscillation cycles.

- The large temporal coherence, quantified with a large coherence time or coherence length, is associated with a narrow spectral bandwidth (or linewidth). For a visible laser beam, this means that it has a certain pure Colour, e.g. red, green or blue, but not white or magenta. Some lasers allow a degree of wavelength tuning (e.g. dye lasers). The large coherence length introduces a tendency for the phenomenon of laser speckle, i.e. a characteristic granular pattern which can be observed e.g. when the laser beam hits a metallic surface.
- In most cases, laser light is linearly polarized. This means that the electric field oscillates in a particular spatial direction (→*polarization of laser emission*).

Depending on the case, laser light can have other remarkable properties:

- Laser light may be visible, but most lasers actually emit in other spectral regions, in particular in the near-infrared region, which human eyes cannot perceive.
- Laser light is not always continuous, but may be delivered in the form of short or ultrashort pulses. As a consequence, the peak power can be extremely high – for some amplified laser systems well above 1 TW (10^{12} W) – if the pulse duration is far below the temporal distance of the pulses, so that a large pulse energy is possible.
- The noise properties of lasers can also be very interesting (→*laser noise*). For example, the oscillation frequency of a laser can be stabilized to stay within an extremely narrow range.

For comparison, light from an incandescent lamp is partly visible and mostly infrared, has a very high spectral bandwidth, cannot be strongly focused (because of low spatial coherence), and cannot be generated in the form of short pulses.

Laser light can cause significant hazards, particularly for the eye, but also for the skin, apart from fire hazards and others. Lasers pulses are often particularly dangerous, because they can have enormously high optical intensities, and invisible laser beams add specific risks..

MANUFACTURING

Lasers are widely used in manufacturing, e.g. for cutting, drilling, welding, cladding, soldering, hardening, ablating, surface treatment, marking, engraving, micromachining, pulsed laser deposition, lithography, alignment, etc. In most cases, relatively high optical intensities are applied to a small spot, leading to intense heating, possibly evaporation and plasma generation. Essential aspects are the high spatial coherence of laser light, allowing for strong focusing, and often also the potential for generating intense pulses.

Laser processing methods have many advantages, compared with mechanical approaches. They allow the fabrication of very fine structures with high quality, avoiding mechanical stress such as caused by mechanical drills and blades. A laser beam with high beam quality can be used to drill very fine and deep holes, e.g. for injection nozzles. A high processing speed is often achieved, e.g. in the fabrication of filter sieves. Further, the lifetime limitation of mechanical tools is removed. It can also be advantageous to process materials without touching them.

Medical Applications

There is a wide range of medical applications. Often these relate to the outer parts of the human body, which are easily reached with light; examples are eye surgery and vision correction (LASIK), dentistry, dermatology (e.g. photodynamic therapy of cancer), and various kinds of cosmetic treatment such as tattoo removal and hair removal.

Lasers are also used for surgery (e.g. of the prostate), exploiting the possibility to cut tissues while causing minimal bleeding.

Very different types of lasers are required for medical applications, depending on the optical wavelength, output power, pulse format, etc. In many cases, the laser wavelength is chosen such that certain substances (e.g. pigments in tattoos or caries in teeth) absorb light more strongly than surrounding tissue, so that they can be more precisely targeted.

Medical lasers are not always used for therapy. Some of them rather assist the diagnosis, e.g. via methods of laser microscopy or spectroscopy.

Metrology

Lasers are widely used in optical metrology, e.g. for extremely precise position measurements with interferometers, for long-distance range finding and navigation. Laser scanners are based on collimated laser beams, which can read e.g. bar codes or other graphics over some distance. It is also possible to scan three-dimensional objects, e.g. in the context of crime scene investigation (CSI).

Optical sampling is a technique applied for the characterization of fast electronic microcircuits, microwave photonics, terahertz science, etc.

Lasers also allow for extremely precise time measurements and are therefore essential component of optical clocks which are beginning to outperform the currently used cesium atomic clocks. Fibre-optic sensors, often probed with laser light, allow for the distributed measurement of temperature, stress, and other quantities e.g. in oil pipelines and wings of airplanes.

Data Storage

Optical data storage e.g. in compact disks (CDs), DVDs, Blu-ray Discs and magneto-optical disks, nearly always relies on a laser source, which has a high spatial coherence and can thus be used to address very tiny spots in the recording medium, allowing a very high density data storage. Another case is holography, where the temporal coherence can also be important.

Communications

Optical fibre communication, extensively used particularly for long-distance optical data transmission, mostly relies on laser light in optical glass fibers. Free-space optical communications, e.g. for inter-satellite communications, is based on higher-power lasers, generating collimated laser beams which propagate over large distances with small beam divergence.

Displays

Laser projection displays containing RGB sources can be used for cinemas, home videos, flight simulators, etc., and are often superior to other displays concerning possible screen dimensions, resolution and Colour saturation. However, further reductions in manufacturing costs will be essential for deep market penetration.

Spectroscopy

Laser spectroscopy is useful, e.g. in atmospheric physics and pollution monitoring (e.g. trace gas sensing with differential absorption LIDAR technology). It also plays a role in medicine (e.g. cancer detection), biology, and various types of fundamental research, partly related to metrology.

Microscopy

Laser microscopes and setups for coherence tomography provide images of, e.g., biological samples with very high resolution, often in three dimensions. It is also possible to realise functional imaging.

Various Scientific Applications

Laser cooling makes it possible to bring clouds of atoms or ions to extremely low temperatures. This has applications in fundamental research and also for industrial purposes.

Particularly in biological and medical research, optical tweezers can be used for trapping and manipulating small particles, such as bacteria or parts of living cells.

Laser guide stars are used in astronomical observatories in combination with adaptive optics for atmospheric correction. They allow substantially increased image resolution even in cases where a sufficiently close-by natural guide star is not available.

Energy Technology

In the future, high-power laser systems might play a role in electricity generation. Laser-induced nuclear fusion is investigated as a alternative to other types of fusion reactors. High-power lasers can also be used for isotope separation.

Military Applications

There are a variety of military laser applications. In relatively few cases, lasers are used as weapons; the "laser sword" has become popular in movies, but not in practice. Some high-power lasers are currently developed for potential use as *directed energy weapons* on the battle field, or for destroying missiles, projectiles and mines.

In other cases, lasers function as target designators or laser sights (essentially laser pointers emitting visible or invisible laser beams), or as irritating or blinding (normally not directly destroying) countermeasures e.g. against heat-seeking anti-aircraft missiles. It is also possible to blind soldiers temporarily or permanently with laser beams, although the latter is forbidden by rules of war.

There are also many laser applications which are not specific for military use, e.g. in areas such as range finding, LIDAR, and optical communications.

Laser Design

The term *design* can have two different meanings. In some cases, it is meant to be a detailed description of a device, including e.g. used parts, how they are put together, and important operation parameters. In other cases, the term denotes the process leading to such a description.

Defining the Design Goals

Before a design is made, the *design goals* must be carefully evaluated. These should include not only the central performance parameters such as output power and wavelength; many more details can be relevant:

*In Short:*Details of the design of a laser product can have a strong impact on its performance, reliability, flexibility, and manufacturing cost.

It is essential to be aware of all the design goals, to know the design properties required for reaching them, and to use a proper laser design (which is worked out at the beginning) as a vital part of an efficient development process.

- Optimum performance, e.g. in terms of output power, power efficiency, beam quality, brightness, intensity and/or phase noise, long-term stability (e.g. of the output power or the optical frequency), timing jitter, etc.
- Compact and convenient setup, ease of operation (e.g. simple turn-on procedure, simple wavelength tuning, no need for realignment)
- Maximum flexibility (e.g. for changing operation parameters)
- Reliability, low maintenance requirements, simple and cost-effective error analysis, maintenance and repair
- Minimum sensitivity to vibrations, temperature changes, electromagnetic interference, aging of components
- Low production cost, i.e., a small number of parts, simple alignment and testing, avoiding the use of parts which are expensive, sensitive, or difficult to obtain

It is certainly advisable to work out carefully the list of these requirements for the particular case before investing any significant resources in laser development, because it can easily be much more expensive and time-consuming to introduce additional properties into an already existing device.

IMPORTANT ASPECTS OF LASER DESIGNS

The properties of the designed laser device are largely determined by the design details, not only by the parts used. Some aspects are particularly important:

- General design parameters, such as resonator length (influencing compactness, tuning issues, frequency stability, etc.), pump intensity
- Selection of gain medium (e.g. a laser crystal) and pump source, suitable choice of geometry (e.g. rod or thin disk, side pumping or end pumping), doping concentration, crystal length, etc.
- Pump setup (e.g. for diode-pumped lasers), influencing output power and beam quality, long-term stability, and the ease of exchanging pump diodes
- Optimum type of laser resonator (e.g. as linear or ring laser, with monolithic or with discrete elements) and optimized resonator design, influencing aspects such as the number of parts, the output power and beam quality, alignment tolerances, sensitivity to thermal lensing, mechanical stability and drifts

- Selection and placement of laser mirrors and intracavity components for wavelength tuning, generation of short pulses via mode locking, dispersion compensation, frequency stabilization, etc.
- Mechanical housing, influencing mechanical stability, efficiency of cooling, temperature drifts, ease of maintenance, and safety issues
- Electronic equipment, e.g. for stabilizing the output power, controlling the laser wavelength, monitoring the status of pump diodes or temperatures, ensuring safe operation
- Proper documentation, including a part list (possibly with suppliers), mechanical designs, alignment and testing procedures, design ideas, possibly optional extensions and limitations for modifying operation parameters

This list, which is certainly not yet complete, shows that proper laser designs are not a trivial matter, but are essential for achieving full customer satisfaction, cost efficiency, and flexibility for future developments.

Needed for Designing Lasers

Designing a laser is a challenging task. The following are definitely required:

- A detailed understanding of the requirements, which may include some understanding of the application
- A detailed understanding of all the relevant physical effects, such as laser amplification, thermal lensing, resonator modes, laser noise, etc., and their interaction
- The essential data, e.g. of laser crystals
- The ability to reduce the complexity to a practical level without losing important details
- Flexible software for calculations and simulations
- Practical experience with lasers, enabling one to recognize typical problems, correctly interpret experimental observations, etc.

It is clear that software alone is by no means sufficient to work out good laser designs.

Role of a Design in a Development Project

It is common practice, but nevertheless generally not advisable, to consider a laser design as a result of a development process which is largely based on trial and error. The design then plays a minor role, just summarizing the results of a lengthy process.

In such cases, the design is often not even properly documented, which creates a risk of losing a lot of potentially valuable information while saving only a minor amount of time at the moment.

In any non-trivial design project – and laser design projects are hardly ever trivial – it is very advisable to attribute a vital role to the laser design:

- The laser design *is made in the office,* not in the laboratory, and properly deals with all known issues which can be or become relevant. Although this process can be much faster and cheaper than a trial-and-error approach in the lab, it takes some considerable discipline and of course requires a comprehensive expertise.
- A proper design is not just a set of ideas, but a very specific description, including e.g. the list of required parts, a more or less detailed prescription on how to put them together wherever this is not trivial, and is ideally supplemented by a description of the underlying reasoning, a discussion of limitations, etc.
- The prototype is then fabricated according to the design, and not vice versa. This greatly speeds up the fabrication, thus making efficient use of costly laboratory resources.
- For any future development of similar kind, the carefully worked out design will be a very valuable input. If it does not exist, and particularly if in addition a vital person has left the company, future developments will be much less efficient.

Attempts to abbreviate this process carry the risk of obtaining reduced performance values and of large time delays due to unexpected technical problems. The later such problems are recognized, understood and solved, the larger can be the resulting damage.

Deriving Designs from Older Designs

In industrial development, it is common to derive some product design from an older design, rather than starting from scratch.

Although this appears very economical, there are significant risks, particularly in cases where the first design has not be properly worked out and documented in a process as described above. A central challenge is that modifying some detail of a laser design may easily have unexpected side effects, introducing new problems which then require additional measures, which again can have side effects.

For such reasons, starting with some initial design, which works e.g. with some lower than desired output power, can be helpful, but it still requires a detailed understanding of that design and its limitations. A proper design document for the initial design can make it easy to produce a whole family of designs, which differ in e.g. output power or pulse repetition rate.

LASER RESONATORS

A laser requires a *laser resonator* (or *laser cavity*), in which the laser radiation can circulate and pass a gain medium which compensates the optical losses. Exceptions are only exotic cases where a medium with very high gain is used, so that amplified spontaneous emission extracts significant power in a single pass through the gain medium.

The laser radiation is automatically generated at one or several frequencies corresponding to resonances (resonator modes), possibly with small deviations caused by "gain pulling". No special measures are required for operating on the resonance; this is different for external resonators, e.g. resonant enhancement cavities.

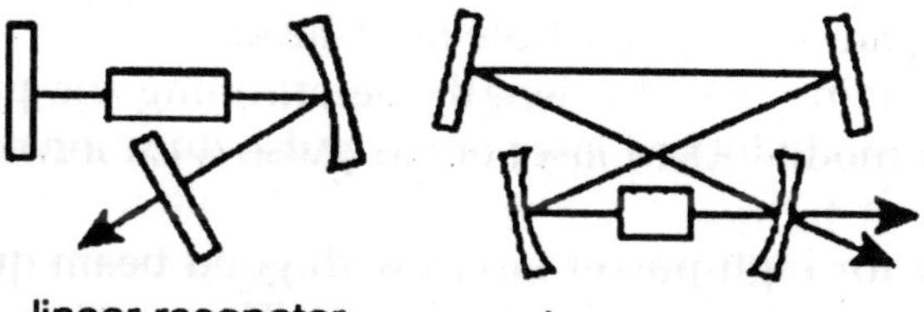

Fig. Two Simple Solid-State Laser resonators with a Laser Crystal as Gain Medium.

Output beams are generated where resonator mirrors are partially transmissive. The ring laser (right) can exhibit laser action in two directions, thus generate two output beams; unidirectional operation could be enforced, e.g. with an additional intracavity optical isolator.

LASER RESONATORS OF SOLID-STATE LASERS

Solid-state bulk lasers are usually built with several dielectric mirrors (laser mirrors), which may be plain or curved. A linear resonator and a ring resonator built in that way, and containing a laser crystal as the gain medium. In some cases, a dielectric mirror coating is placed on the gain medium itself on monolithic solid-state lasers. One of the mirrors, usually an end mirror, is the partially transmissive output coupler.

The design of the laser resonator (comprising optical elements, angles of incidence, and distances between the components) determines the beam radius of the fundamental mode at all locations along the beam, together with other important properties. For maximum beam quality (→ *diffraction-limited* output), the beam radius in the gain medium has to match approximately the radius of the pumped region.

For smaller beam radii, operation with multiple spatial modes is obtained, leading to a non-ideal beam quality; however, such multimode lasers have other advantages such as much wider stability zones and a lower sensitivity to misalignment.

In many cases, the laser resonator design should have additional features.

For example, it can be optimized:

- For compactness
- For achieving certain values of the beam radius in other optical components (e.g. on a saturable absorber of a passively mode-locked laser)
- For avoiding too small beam radii in optical components (leading to optical damage particularly in Q-switched lasers)
- For minimizing adverse effects of thermal lensing and aberrations in the gain medium
- For minimizing the alignment sensitivity
- For accommodating multiple laser heads
- For a certain resonator length, determining the pulse repetition rate in a mode-locked laser or the pulse duration of a Q-switched laser.

Particularly for high-power lasers with good beam quality, thermal lensing in the gain medium is very important. The resonator design should be made so that changes of the thermal lens do not affect too much the mode sizes.

Also, it should have a low sensitivity to thermal aberrations and misalignment. The importance of these factors should not be underestimated; there are cases where two resonators even with equal mode sizes in the gain medium lead to very different laser performance and are radically different in terms of alignment.

Although it is normally not that difficult to evaluate the properties of a given laser resonator, it can be challenging to find a resonator design which satisfies multiple criteria such as those listed above.

Numerical optimization, using special resonator design software, can be the only way to find good solutions, particularly for some mode-locked lasers. Also, a solid understanding of resonator properties can help considerably when trying to find resonator configurations with special combinations of properties, such as large mode areas and short lengths. For advanced design issues, a great deal of experience is at least as important as a versatile design software.

Some high-power lasers (for example with slab designs) are operated with *unstable resonators,* allowing a reasonable (but typically not diffraction-limited) beam quality to be achieved despite the presence of strong thermal effects in the gain medium. Due to the high diffraction losses, such laser cavities require relatively high gain.

There are various types of monolithic solid-state lasers which have the whole beam path within the laser crystal. Beam reflections are then typically realized either with dielectric coatings on crystal surfaces, or with total internal reflection.

Laser Crystals

Laser crystals are typically single crystals (monocrystalline material) which are used as gain media for solid-state lasers. In most cases, they are doped with either trivalent rare earth ions or transition metal ions. These ions enable the crystal to amplify light at the laser wavelength via stimulated emission, when energy is supplied to the crystal via absorption of pump light (→ *optical pumping*). Compared with doped glasses, crystals usually have higher transition cross sections, a smaller absorption and emission bandwidth, a higher thermal conductivity, and possibly birefringence. In some cases, monocrystalline laser materials may be replaced with ceramic gain media, which have a fine polycrystalline structure.

Common Laser-active Dopants

The most frequently used laser-active rare earth ions and host media together with some typical emission wavelengths:

Table. Common rare Earth Ions in Laser–Active Crystals.

Ion	Common host crystals	Important emission wavelengths
Neodymium (Nd^{3+})	$Y_3Al_5O_{12}$ (YAG), $YAlO_3$ (YALO), YVO_4(yttrium vanadate), $YLiF_4$ (YLF), tungstates ($KGd(WO_4)_2$, $KY(WO_4)_2$)	1064, 1047, 1053, 1342, 946 nm
Ytterbium (Yb^{3+})	YAG, tungstates (e.g. KGW, KYW, KLuW), YVO_4, borates (BOYS, GdCOB), apatites (SYS), sesquioxides (Y_2O_3, Sc_2O_3) nm	1030, 1020–1070
Erbium (Er^{3+})	YAG, YLF	2.9, 1.6 μm
Thulium (Tm^{3+})	YAG	1.9–2.1 μm
Holmium (Ho^{3+})	YAG	2.1, 2.94 μm
Cerium (Ce^{3+})	YLF, LiCAF, LiLuF, LiSAF, and similar fluorides	0.28–0.33 μm

The following tables lists common transition-metal doped crystals:

Table. Common Transition Metal ions in Laser–Active Crystals.

Ion	Common host crystals	Important emission wavelengths
Titanium (Ti^{3+})	sapphire	650–1100 nm
Chromium (II) (Cr^{2+})	zinc chalcogenides such as ZnS, ZnSe, and ZnS_xSe_{1-x}	2–3.4 μm
Chromium (III) (Cr^{3+})	Al_2O_3 (ruby), $LiSrAlF_6$ (LiSAF), LiC aAlF$_6$ (LiCAF), $LiSrGaF_6$ (LiSGAF)	0.8–0.9 μm
Chromium (IV)	YAG, $MgSiO_4$ (forsterite)	1.35–1.65 μm (YAG), 1.1–1.37 μm forsterite) (Cr^{4+})

These tables contain only the most common host crystals; many others exist, which are less frequently used.

Important Properties of Host Crystals

The host crystal is much more than just a means to fix the laser-active ions at certain positions in space. A number of properties of the host material are important:

- The medium should have a high transparency (low absorption and scattering) in the wavelength regions of pump and laser radiation, and good optical homogeneity. To some extent, this depends on the quality of the material, determined by details of the fabrication process.
- The host medium influences strongly the wavelength, bandwidth and transition cross sections of pump and laser transitions and also the upper-state lifetime. For example, Nd:YVO_4 has much higher cross sections, a larger gain bandwidth, and a smaller upper-state lifetime than Nd:YAG. Other neodymium hosts provide other transition wavelengths, e.g. 1047 or 1053 nm from Nd:YLF.
- Nonradiative transitions (e.g. multi-phonon transitions) are also strongly influenced by the host, in particular by its maximum phonon energy. Some of these transitions are very detrimental, leading to quenching of the upper-state population (thus lowering the quantum efficiency). Others are essential for laser operation, e.g. for removing ions from the lower laser level.
  Energy transfer processes are also dependent on the host material.
- The maximum possible doping concentration can depend strongly on the host material and its fabrication method.
- Different crystalline materials are very different concerning their hardness and other properties, which determine with which methods and how easily they can be cut and polished with good quality.
- Some materials are chemically not stable, e.g. hygroscopic.
- Particularly for high-power lasers (but often enough also for medium and low powers), a high thermal conductivity low thermo-optic coefficients (for weak thermal lensing) and a high resistance to mechanical stress are desirable.
- Optical isotropy can be beneficial, but in other cases birefringence (reducing thermal depolarization) and possibly polarization-dependent gain is preferable.
- A high damage threshold in terms of pulse fluence or peak intensity can be important for high-energy amplifiers.

It is apparent that different applications lead to very different requirements on laser gain media. For this reason, a broad range of different crystals are used, and making the right choice is essential for constructing lasers with optimum performance.

Common Crystalline Laser Host Media

There is a wide range of crystalline media, which can be grouped according to important atomic constituents and crystalline structures. Some important groups of crystals are:

- Garnets such as $Y_3Al_5O_{12}$ (YAG), $Gd_3Ga_5O_{12}$ (GGG), and $Gd_3Sc_2Al_3O_{12}$ (GSGG): hard and chemically inert materials, optical isotropic, with high thermal conductivity
- sapphire (Al_2O_3) (e.g. for titanium–sapphire lasers) and aluminates such as $YAlO_3$ (YALO, YAP) for neodymium doping: high hardness and thermal conductivity, anisotropic
- Sesquioxides such as Y_2O_3, Sc_2O_3: isotropic, high hardness and thermal conductivity
- Vanadates such as YVO_4 and $GdVO_4$: very high laser cross sections of Nd^{3+}, anisotropic
- fluorides, e.g. $YLiF_4$ (YLF): good UV transparency, birefringence, large energy storage capability of Nd:YLF; also LiCAF, LiLuF, LiSAF as chromium-doped broadband gain media
- Silicates, e.g. $MgSiO_4$ (forsterite): broad gain bandwidth
- monoclinic double tungstates such as $KGd(WO_4)_2$ (KGW) and $KY(WO_4)_2$ (KYW): combination of relatively high Yb^{3+} laser cross sections, large gain bandwidth, and high thermal conductivity
- Disordered tetragonal double tungstates such as $NaGd(WO_4)_2$ (NGW) and $NaY(WO_4)_2$ (NYW): particularly large gain bandwidth of ytterbium
- Chalcogenides such as ZnS or ZnSe for mid-infrared lasers

Geometries of Laser Crystals

Different geometric forms can be used in lasers:

- A common form is that of a cuboid. The crystal can be, e.g., a thin coplanar plate, with transverse dimensions (perpendicular to the laser beam) and a thickness of a few millimeters. It may be oriented for near perpendicular incidence of the laser beam, or at Brewster's angle.
- It can be fixed in some solid mount which also acts as a heat sink. Larger crystals are usually used for side pumping e.g. with high-power diode bars.

- In some cases, extreme angles between the end faces are required, e.g. if one end face has to be Brewster-angled while the other one is made for perpendicular incidence.
- Slab lasers are based on relatively flat slabs, which may or may not be of cuboid form.
- Many side-pumped lasers use relatively long cylindrical laser rods, e.g. made of Nd:YAG. Particularly for lamp-pumped lasers, the rod length can be several centimeters, whereas the rod diameter is much smaller (a few millimeters).
- Thin-disk lasers require a disk, often with circular cross section, having a thickness of only e.g. 100–200 ìm and a relatively high doping concentration.
- Special geometries are required for monolithic solid-state lasers, such as nonplanar ring oscillators.
- For various reasons, composite crystals are becoming popular. These have a spatially varying chemical composition and can be made with special shapes.

Bulk Properties

For a given dopant and host medium, the doping concentration is the most important parameter. Other issues of interest are the uniformity of doping (influencing the tendency for quenching), the level of impurities (e.g. unwanted other rare earth ions), and the optical homogeneities. Several of these factors influence the absorption and scattering losses of the material, and/or the strength of thermal lensing.

Of course, it is very desirable that a given crystal quality is produced consistently, although different laser designs can have a different sensitivity to material parameters.

Optimization of Geometry and Parameters

Which geometry, dopant and doping concentration of the gain medium are most advantageous depend on several factors. The available pump source (type of laser diode or lamp) and the envisaged pumping arrangement are important factors, but the material itself also has some influence.

For example, titanium–sapphire lasers have to be pumped with high intensities, for which the form of a transversely cooled rod, operated with relatively small pump and laser beam diameter, is more appropriate than e.g. a thin disk. As another example, Q-switched lasers reach a higher population density in the upper laser level and are therefore more sensitive to quenching effects and energy transfer processes; therefore, a lower doping density is often appropriate for these devices.

For high-power lasers, lower doping densities are often used in order to limit the density of heat generation, although thin-disk lasers work best with highly doped crystals. Many laser products do not reach the full performance potential because such details have not been properly worked out.

Optical Surfaces

Those surfaces which are passed by the laser beam are normally either oriented at Brewster's angle or have an anti-reflection coating. Even AR-coated crystals are often slightly tilted against the beam so as to prevent back-reflections staying in the laser resonator. This is important for, e.g., mode-locked lasers and tunable single-frequency lasers.

A high surface quality is of course important. Specifications of surface flatness are often better than $\lambda/10$. This helps to avoid both scattering losses and wavefront distortions which can degrade the laser's beam quality. In addition, *scratch and dig specifications* (*cosmetic surface quality*) limit the density of small-scale surface defects; they may read e.g. "80–50" for medium quality mass production, or "10–5" for particularly demanding laser applications.

Proper surface treatment also influences the damage threshold, which is important e.g. for high-energy pulse amplifiers. Finally, a high degree of end face parallelism can be important for avoiding changes of beam direction in a crystal.

Diode-pumped Lasers

Virtually all optically pumped lasers fall into one of two categories:

- *Lamp-pumped lasers,* having some kind of discharge lamp as the pump source
- *Diode-pumped lasers,* pumped with some kind of laser diodes

The latter category, for which the term all-solid-state lasers is also used.

Types of Diode-pumped Lasers

Most diode-pumped lasers are solid-state lasers (DPSSL = DPSS lasers = *diode-pumped solid-state lasers*). These are either bulk lasers, using some kind of laser crystal or bulk piece of glass, or fibre lasers (although the term DPSSL is less common for fibre lasers). Both categories span a range of output powers from a few milliwatts to multiple kilowatts (→ *high-power lasers*).

Less common are optically pumped semiconductor lasers (particularly VECSELs = vertical external cavity surface-emitting lasers), and there are also some relatively exotic types of diode-pumped gas lasers, e.g. alkali vapour lasers.

Types of Laser Diodes for Diode Pumping

There are different types of laser diodes which can be used for diode pumping:

- Low-power lasers (up to roughly 200 mW) can be pumped with small edge-emitting laser diodes. These exhibit a diffraction-limited beam quality and make it fairly easy to achieve the same for the solid-state laser.
- Broad area laser diodes typically generate several watts and are suitable for pumping solid-state lasers with output powers up to a few watts. Their beam quality is substantially asymmetric, but normally still sufficient for achieving a diffraction-limited laser output without using complicated optics.
- High-power diode bars emit tens of watts (or even > 100 W), allowing for higher output powers, particularly when several bars are combined. Their beam quality is strongly asymmetric and fairly poor, so that their brightness (radiance) is much lower than that of lower-power diodes. Various types of beam shapers are used to symmetrize the beam quality. This makes it easier either to pump a bulk laser or to couple the light into a fibre.
- For the highest powers, diode stacks are often used. These have a still worse beam quality and lower brightness, but can provide multiple kilowatts.

In most cases, the pump diodes are operated continuously. This applies to all continuous-wave and mode-locked lasers, and also to many Q-switched lasers. However, quasi-continuous-wave operation with higher peak power for limited time intervals (e.g. 100 ìs) is sometimes used for Q-switched lasers with a high pulse energy and low pulse repetition rate.

Depending on the type of laser diode, different kinds of pump optics are used. It is also possible to use fibre-coupled diode lasers, which make it possible to separate the actual laser head from another package containing the pump diodes, so that the laser head can become very compact.

Advantages of Diode Pumping

The main advantages of diode pumping are:

- A high electrical-to-optical efficiency of the pump source (of the order of 50 %) leads to a high overall power efficiency of the laser. As a consequence, small power supplies are needed, and both the electricity consumption and the cooling demands are drastically reduced, compared with those for lamp-pumped lasers.
- The narrow optical bandwidth of diode lasers makes it possible to pump directly certain transitions of laser-active ions without

losing power in other spectral regions. It thus also contributes to a high efficiency.

- Although the beam quality of high-power diode lasers is not perfect, it often allows for end pumping of lasers with very good overlap of laser mode and pump region, leading to high beam quality and power efficiency. In the domain of slab lasers, it allows edge pumping instead of face pumping, which brings important advantages.
- Diode-pumped low-power lasers can be pumped with diffraction-limited laser diodes. This allows the construction of very low-power lasers with reasonable power efficiency, i.e., with a very small electrical pump power, as is important for battery-powered devices.
- The lifetime of laser diodes is long compared with that of discharge lamps: typically many thousands of hours, often even well above 10 000 hours. However, various factors can lead to drastically reduced lifetimes. Also, the exchange of laser diodes is much more costly than that of discharge lamps.
- The compactness of the pump source, the power supply and the cooling arrangement makes the whole laser system much smaller and easier to use.
- Diode pumping makes it possible to use a very wide range of solid-state gain media for different wavelength regions, including e.g. upconversion lasers. For many solid-state gain media, the lower brightness of discharge lamps would not be sufficient.
- The low intensity noise of laser diodes leads to low noise of the diode-pumped laser.

Achievements

The benefits of diode pumping have lead to amazing achievements. Some examples are:

- There are miniature solid-state lasers with excellent efficiency, beam quality, spectral purity, and stability. Some of those may even be battery-powered.
- Diode-pumped high-power solid-state lasers can deliver kilowatt output powers with fairly high beam quality. This applies particularly to thin-disk lasers, but also to high-power fibre lasers and amplifiers.
- Diode pumping is also essential for a large variety of mode-locked lasers, generating e.g. up to 80 W of average output power in sub-picosecond pulses, or picosecond pulses for telecom applications with a pulse repetition rate up to 50 GHz.

Limitations

In the early years of diode pumping, the output powers achievable were very limited – smaller than those of lamp-pumped lasers. In the meantime, however, high-power diode bars and diode stacks have become very powerful, and the highest output powers are now usually achieved with diode pumping.

The main disadvantage of diode pumping (as compared with lamp pumping) is the significantly higher cost per watt of pump power. This is severe for high powers. For this reason, lamp pumping is still used in cases where high powers are needed, particularly when the power is used only for short times.

For example, lamp-pumped Q-switched Nd:YAG lasers are still widely used for laser marking, and will not soon be replaced with diode-pumped lasers.

Laser diodes are electrically less robust than discharge lamps. They may e.g. be quickly destroyed by excessive drive currents, or by electrostatic discharges. In conjunction with properly designed electronics, however, this should not happen. Problems can also arise from optical feedback.

Applications

Diode-pumped solid-state lasers have a very wide range of applications. Indeed, they are used in all of the areas mentioned in the article on laser applications.

FIBRE LASERS

Fibre lasers are usually meant to be lasers with optical fibers as gain media, although some lasers with a semiconductor gain medium (a semiconductor optical amplifier) and a fibre resonator have also been called fibre lasers (or *semiconductor fibre lasers*). In most cases, the gain medium is a fibre doped with rare earth ions such as erbium (Er^{3+}), neodymium (Nd^{3+}), ytterbium (Yb^{3+}), thulium (Tm^{3+}), or praseodymium (Pr^{3+}), and one or several laser diodes are used for pumping (→ *diode pumping*).

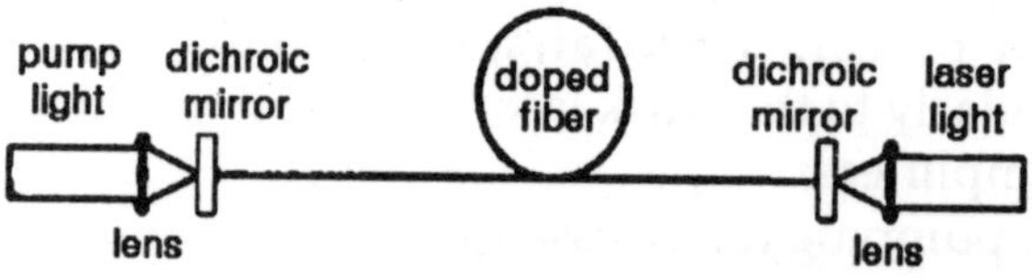

Fig. Setup of A Simple Fibre Laser. Pump Light is Launched from the Left-Hand Side Through a Dichroic Mirror into the Core of the Doped Fibre. The Generated Laser Light is Extracted on The Right-Hand Side.

FIBRE LASER RESONATORS

In order to form a laser resonator with fibers, one either needs some kind of reflector (mirror) to form a linear resonator, or one builds a fibre ring laser. Various types of mirrors are used in linear fibre laser resonators:

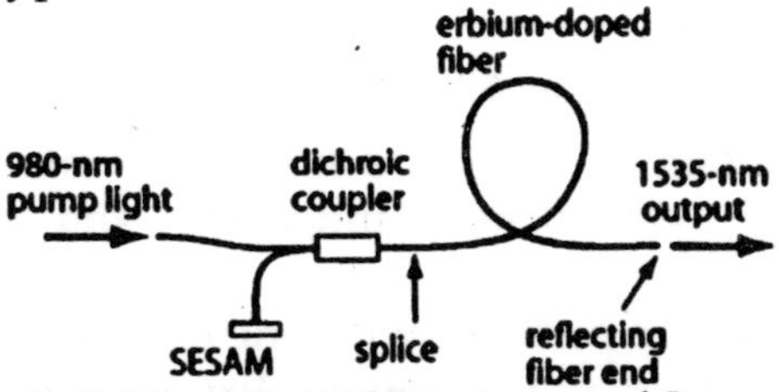

Fig. A Simple Erbium-Doped Femtosecond Laser, where the Fresnel Reflection from a Fibre end is used for Output Coupling.

- In simple laboratory setups, ordinary dielectric mirrors can be butted to the perpendicularly cleaved fibre ends. This approach, however, is not very practical for mass fabrication and not very durable either.
- The Fresnel reflection from a bare fibre end face is often sufficient for the output coupler of a fibre laser.
- It is also possible to deposit dielectric coatings directly on fibre ends, usually with some evaporation method. Such coatings can be used to realise reflectivities in a wide range.
- For commercial products, it is common to use fibre Bragg gratings, made either directly in the doped fibre, or in an undoped fibre which is spliced to the active fibre.

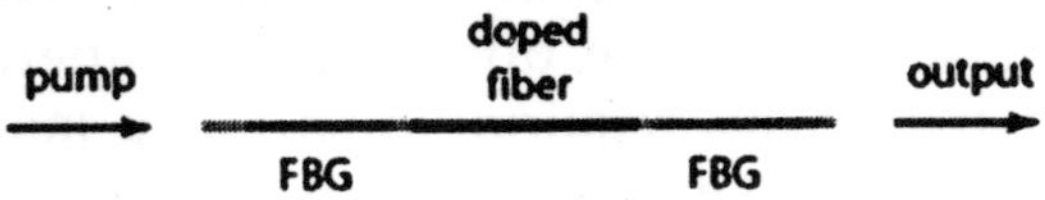

Fig. Short DBR Fibre Laser for Narrow–Linewidth Emission.

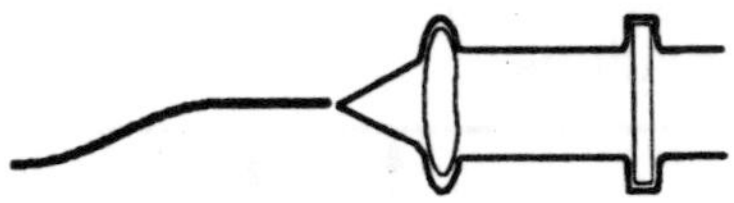

Fig. End Reflector with Lens and Mirror.

- A better power-handling capability is achieved by collimating the light exiting the fibre with a lens and reflecting it back with a dielectric mirror. The intensities on the mirror are then greatly reduced due to the much larger beam area. However, slight misalignment can cause substantial reflection losses, and the additional Fresnel reflection at the fibre end can introduce filter effects and the like. The latter effects can be suppressed by using angle-cleaved fibre ends, which however introduce polarization-dependent losses.

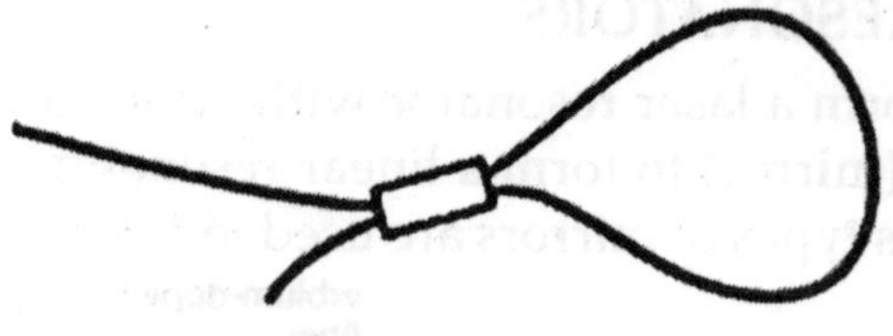

Fig. Fibre Loop Mirror.

- Another option is to form a fibre loop mirror, based on a fibre coupler (e.g. with 50:50 splitting ratio) and some piece of passive fibre.

Most fibre lasers are pumped with one or several fibre-coupled diode lasers. The pump light may be coupled directly into the core, or in high-power into a larger pump cladding (→ *double-clad fibers*). There are many different kinds of fibre lasers, some of which are discussed in the following.

High-power Fibre Lasers

Whereas the first fibre lasers could deliver only a few milliwatts of output power, there are now high-power fibre lasers with output powers of hundreds of watts, sometimes even several kilowatts from a single fibre. This potential arises from a high surface-to-volume ratio (avoiding excessive heating) and the guiding effect, which avoids thermo-optical problems even under conditions of significant heating.

Upconversion Fibre Lasers

The fibre laser concept is most suitable for the realization of upconversion lasers, as these often have to operate on relatively "difficult" laser transitions, requiring high pump intensities. In a fibre laser, such high pump intensities can be easily maintained over a long length, so that the gain efficiency achievable often makes it easy to operate even on low-gain transitions.

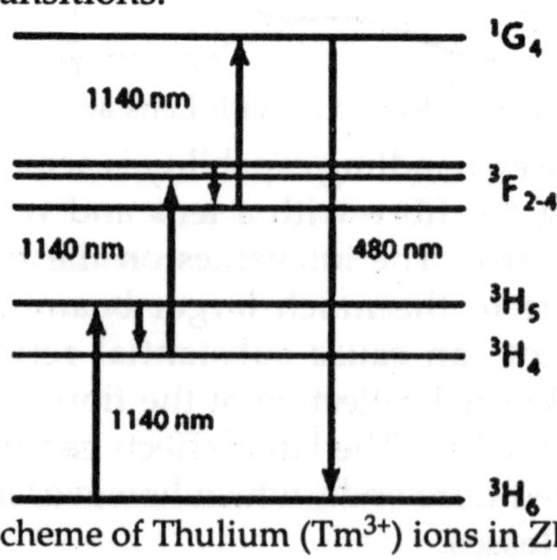

Fig. Level Scheme of Thulium (Tm^{3+}) ions in ZBLAN Fibre.

In most cases, silica glass is not suitable for upconversion fibre lasers, because the upconversion scheme requires relatively long lifetimes of intermediate electronic levels, and such lifetimes are often very small in silica fibers due to the relatively large phonon energy of silica glass (→ *multi-phonon transitions*). Therefore, one mostly uses certain heavy-metal fluoride fibers such as ZBLAN (a fluorozirconate) with low phonon energies.

The probably most popular upconversion fibre lasers are based on thulium-doped fibers for blue light generation, praseodymium-doped lasers (possibly with ytterbium codoping) for red, orange, green or blue output, and green erbium-doped lasers.

Narrow-linewidth Fibre Lasers

Fibre lasers can be constructed to operate on a single longitudinal mode (→ *single-frequency lasers, single-mode operation*) with a very narrow linewidth of a few kilohertz or even below 1 kHz. In order to achieve long-term stable single-frequency operation without excessive requirements concerning temperature stability, one usually has to keep the laser resonator relatively short (e.g. of the order of 5 cm), even though a longer resonator would in principle allow for even lower phase noise and a correspondingly smaller linewidth.

The fibre ends have narrow-bandwidth fibre Bragg gratings (→ *distributed Bragg reflector lasers, DBR fibre lasers*), selecting a single resonator mode. Typical output powers are a few milliwatts to some tens of milliwatts, although single-frequency fibre lasers with up to roughly 1 W output power have also been demonstrated.

An extreme form is the distributed-feedback laser (DFB laser), where the whole laser resonator is contained in a fibre Bragg grating with a phase shift in the middle. Here, the resonator is fairly short, which can compromise the output power and linewidth, but single-frequency operation is very stable.

Of course, further amplification to much higher power levels in a fibre amplifier is possible.

Q-switched Fibre Lasers

With various methods of active or passive Q switching, fibre lasers can be used for generating pulses with durations which are typically between tens and hundreds of nanoseconds. The pulse energy achievable with large mode area fibers can be several millijoules, in extreme cases tens of millijoules, and is essentially limited by the saturation energy (even for large mode area fibers) and by the damage threshold (the latter particularly for shorter pulses).

As fibre lasers typically have relatively long resonators (particularly for high-power lasers based on double-clad fibers), the pulse durations tend to be longer than those of bulk lasers.

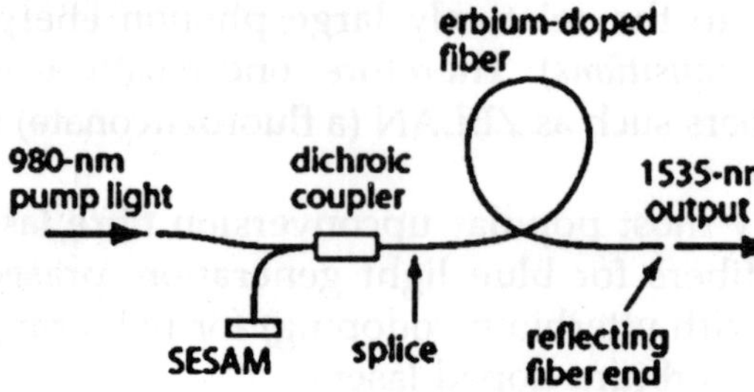

Fig. Simple Erbium-Doped Q-Switched Fibre Laser.

Mode-locked Fibre Lasers

More sophisticated resonator setups are used particularly for mode-locked fibre lasers (*ultrafast fibre lasers*), generating picosecond or femtosecond pulses. Here, the laser resonator may contain an active modulator or some kind of saturable absorber.

An artificial saturable absorber can be constructed using the effect of nonlinear polarization rotation, or a nonlinear fibre loop mirror. A nonlinear loop mirror is used e.g. in a "figure-of-eight laser" where there is a main resonator on the left-hand side and a nonlinear fibre loop, which does the amplification, shaping and stabilization of a circulating ultrashort pulse. Particularly for harmonic mode locking, additional means may be used, such as subcavities acting as optical filters.

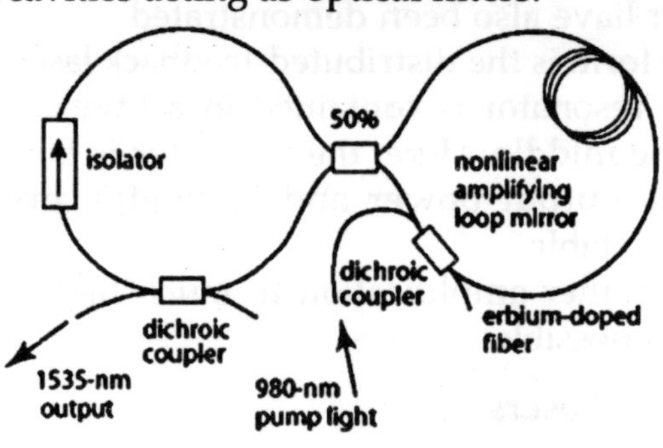

Fig. Figure-of-Eight Laser

RAMAN FIBRE LASERS

A special type of fibre lasers are fibre Raman lasers, relying on Raman gain associated with the fibre nonlinearity. Such lasers usually use relatively long fibers, sometimes of a type with increased nonlinearity, and typical pump powers of the order of 1W.

With several nested pairs of fibre Bragg gratings, the Raman conversion can be done in several steps, bridging hundreds of nanometers between the pump and output wavelength. Raman fibre lasers can e.g. be pumped in the 1-μm region and generate 1.4-μm light as required for pumping 1.5-μm erbium-doped fibre amplifiers.

Fibre Lasers with Semiconductor Optical Amplifiers

There are some lasers which have a semiconductor optical amplifier (SOA) as the gain medium in a resonator made of fibers.

Even though the actual laser process does not occur in a fibre, such fibers are sometimes called fibre lasers. They typically emit relatively small optical powers of a few milliwatts or even less.

Sometimes they exploit the very different properties of the semiconductor gain medium, as compared with a rare-earth-doped fibre, in particular the much smaller saturation energy and upper-state lifetime.

Rather than only generating coherent light, such lasers can be used for information processing in optical fibre communications systems – for example the wave-length conversion of data channels based on cross-saturation effects.

Special Attractions of Fibers as Laser Gain Media

- As fibers can be coiled and the light propagating in fibers is well shielded from the environment (e.g. concerning dust), fibre lasers can have a compact and rugged setup, provided that the whole laser resonator is built only with fibre components (*all-fibre setup*) such as fibre Bragg gratings and fibre couplers (i.e., avoiding free-space optics and any requirement for alignment).
- Fibre gain media have a large gain bandwidth due to strongly broadened laser transitions in glasses, permitting wide wavelength tuning ranges and/or the generation of ultrashort pulses. Also, fibre lasers have broad spectral regions with good pump absorption, making the exact pump wavelength uncritical, so that temperature stabilization of the pump diodes is usually not necessary.
- Diffraction-limited beam quality is easily obtained when single-mode fibers are used, and sometimes also with slightly multimode fibers.
- Due to the high gain efficiency of doped fibers, fibre lasers have the potential to operate with very small pump powers. Also, it is possible to obtain very high power efficiencies.
- In recent years, the potential for very high output powers (several kilowatts with double-clad fibers) has been convincingly demonstrated.

- Again due to the guidance, which allows high pump intensities to be applied over long lengths, fibre lasers can be operated even on very "difficult" laser transitions (e.g. of upconversion lasers).

On the other hand, fibre lasers can suffer from various problems:

- When the pump light has to be launched from free space into a single-mode core, the alignment is critical. This problem can be eliminated by using fibre-coupled pump diodes.
- Most fibers exhibit a complicated temperature-dependent polarization evolution, unless polarization-maintaining fibers or Faraday rotators are used. Such measures, however, are normally not compatible with nonlinear polarization rotation mode locking.
- Nonlinear effects often limit the performance, e.g. in terms of powers achievable in single-frequency operation or the pulse quality of mode-locked lasers. For example, Kelly sidebands are often seen, whereas mode-locked bulk lasers rarely exhibit this effect.
- At high powers, there is a risk of fibre damage even below the actual damage threshold of the material (*fibre fuse*).
- Fibers have a limited gain and pump absorption per unit length, making it difficult to realise very short resonators e.g. for single-frequency lasers or for multi-gigahertz mode-locked lasers. However, significant progress has been made in this direction recently via the development of very highly doped fibers, usually made from phosphate glass.

Double-clad Fibers

A fibre laser or amplifier based on an ordinary doped single-mode fibre can generate a diffraction-limited output, but it restricts the pump sources to those with diffraction-limited beam quality and thus normally to those with low power.

On the other hand, the use of multimode fibers usually (although not always) leads to poor beam quality. This dilemma has been resolved with the invention of double-clad fibre designs, which allow *cladding pumping of fibre devices*. Here, the laser light propagates in a single-mode (or multimode) core, which is surrounded by an *inner cladding* in which the pump light propagates. Only the core (or sometimes a ring around the core) is rare-earth-doped.

The pump light is restricted to the inner cladding by an outer cladding with lower refractive index, and also partly propagates in the single-mode core, where it can be absorbed by the laser-active ions.

The inner cladding has a significantly larger area (compared with that of the core) and typically a much higher numerical aperture, so that it can support a large number of propagation modes, allowing the efficient launch of the output, e.g. of high-power laser diodes (e.g. beam-shaped high-power diode bars), despite their poor beam quality.

The signal light is launched into the doped core, while the pump light is launched into the inner cladding. The core is D-shaped for more efficient pump absorption. The pump light does not necessarily need to be injected into the fibre ends.

It is also possible to use side pumping techniques, where access to the fibre ends is not required for pumping. For example, coated V grooves cut into the inner cladding can be used to reflect pump light into the inner cladding.

Double-clad Fibre Designs

There are a variety of different designs of double-clad fibers. The fibre cross-sections for the most important design types. The simplest kind of design has a circular pump cladding and a centered core.

This is relatively easy to make and use, but in this kind of fibers there are propagation modes of the inner cladding (related to *helical rays*) which have hardly any overlap with the core, so that some significant part of the pump light exhibits incomplete absorption. As a result, the gain and power efficiency are compromised. To a limited extent, this problem can be solved by strongly coiling the fibre.

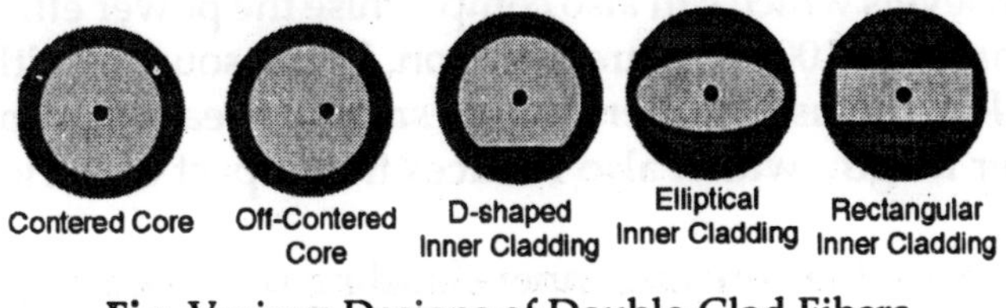

Fig. Various Designs of Double-Clad Fibers.

Modes with poor core overlap can be avoided by using a modified design with a lower symmetry. Examples are designs with an off-centered core or a non-circular (e.g. elliptical, D-shaped or rectangular) inner cladding. Such pump claddings are also often better matching the properties of pump sources such as beam-shaped diode bars.

Here, the multimode pump core is suspended by very thin struts in the air cladding, through which the pump light cannot escape. Such a structure can have a very high numerical aperture of at least 0.6 for the pump light; this further reduces the requirements concerning the brightness of the pump source.

The thickness of the struts can be chosen such that at the same time one achieves good mechanical stability, high thermal conductivity, and minimal pump losses.

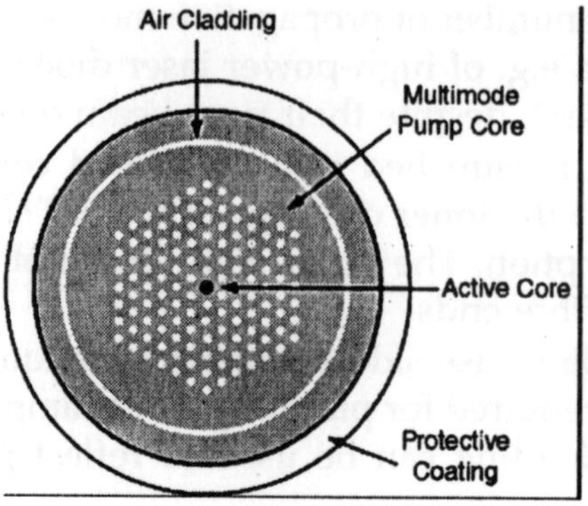

Fig. Structure of a Photonic Crystal Fibre with an Air Cladding.

Another advantage of this type of fibre is that pump light is kept away from the protective polymer coating, avoiding any damage by absorbed pump light. The guidance of the core is achieved as in other photonic crystal fibers.

Parameters and Fabrication Methods of Double-clad Fibers

Besides the properties of the fibre core, the ratio of the areas of inner cladding and core is an important parameter. This area ratio should not be too large, because otherwise the effective pump absorption length becomes large, and the pump intensity in the core is small, resulting in low excitation levels which can also compromise the power efficiency. Area ratios of the order of 100–1000 are common. Pump sources with improved brightness allow the use of fibers with a smaller area ratio, and thus also with a smaller length, which also reduces the impact of various types of nonlinearities.

In many cases, the core and inner cladding of a double-clad fibre are similar to those of a normal core-pumped fibre, except that in addition there is the lower-index outer cladding. If the inner cladding is made of silica, the outer cladding may consist of fluorine-doped silica.

The numerical aperture for the inner cladding can then be e.g. 0.28. Larger values are possible with polymer outer claddings, but these cannot tolerate very high temperatures and may introduce higher propagation losses for the pump light.

Applications

Double-clad fibers are extensively used for cladding-pumped high-power fibre lasers and amplifiers. Such devices can have a fairly high power conversion efficiency combined with good beam quality. As the beam quality of the output can be diffraction-limited whereas that of the pump can be poor, the brightness of the laser or amplifier output can be much higher than that of the pump source. Particularly if this increase in brightness is essential for an application, the cladding-pumped fibre laser may be called a *brightness converter*.

High-power Fibre Lasers and Amplifiers

Whereas the first fibre lasers could deliver only a few milliwatts of output power, there have subsequently been rapid developments which have lead to high-power fibre lasers and particularly amplifiers with output powers of tens or hundreds of watts, sometimes even several kilowatts from a single fibre. This potential arises from a very high surface-to-volume ratio (avoiding excessive heating) and the guiding (waveguide) effect, which avoids thermo-optical problems even under conditions of significant heating. Fibre laser technology now competes strongly with other high-power laser technologies based on solid-state bulk lasers, such as thin-disk lasers.

Double-clad Fibers and Beam Quality

High-power fibre lasers and fibre amplifiers are nearly always realized with rare-earth-doped double-clad fibers, which are pumped with fibre-coupled high-power diode bars or other kinds of laser diodes. The pump light is launched into an inner cladding rather than into the (much smaller) fibre core, in which the laser light is generated.

The laser light can have very good beam quality – even diffraction-limited beam quality if the fibre has a single-mode core. Therefore, the brightness of the fibre laser output can be orders of magnitude higher than that of the pump light, even though the output power is of course somewhat smaller. (Typical power efficiencies are above 50%, sometimes even above 80%.) Such fibre lasers can effectively be used as brightness converters, i.e., devices increasing the brightness.

For the highest powers, the core area needs to be fairly large (→ *large mode area fibers*), because the optical intensities would otherwise become too high, and often also because a double-clad fibre with a large ratio of cladding to core area has a weak pump absorption. For core areas up to the order of a few thousand μm^2, it is feasible to have a single-mode core.

Larger mode areas with still fairly good output beam quality are possible with a slightly multimode core, where most of the light propagates in the fundamental mode. (The excitation of higher-order fibre modes can be suppressed to some extent e.g. by coiling the fibre.)

For even larger mode areas, the beam quality can no longer be nearly diffraction-limited, but it can still be fairly good compared with, e.g., rod lasers operating at similar power levels.

LAUNCHING THE PUMP LIGHT

There are several options for launching pump light at very high power levels. The simplest is to launch directly into the pump cladding at one or both fibre ends. This technique does not require special fibre components; however, it needs the propagation of the high-power pump radiation through air (with free-space optics) and particularly through an air–glass interface, which is then very sensitive to dust and misalignment. In many cases, it is therefore preferable to use one of several techniques where one employs fibre-coupled pump diodes and keeps the pump light in fibers from there on.

One option is to launch the pump light into passive (undoped) fibers which are wound around the active fibre so that the light is gradually transferred into the active fibre (*GTWave fibre*). Other techniques are based on special pump combiner devices, where several pump fibers and a single active signal fibre are fused together. Yet other approaches are based on side-pumped fibre coils (*fibre disk lasers*) or on grooves in the pump cladding through which pump light can be injected. The latter technique allows for *multi-point pump injection* and thus for a better distribution of the heat load.

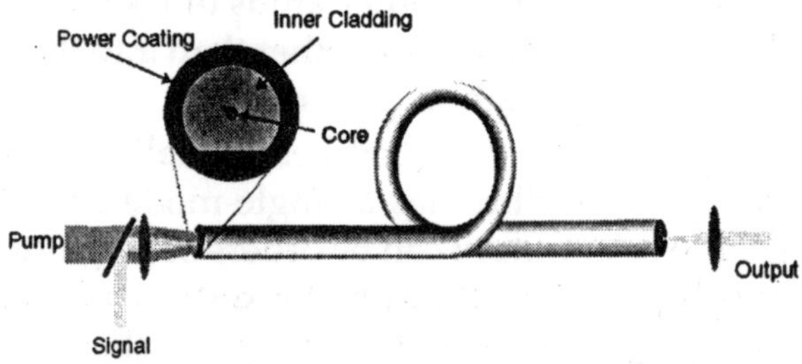

Fig. A High-power Double-Vlad Fibre

A comparison of all the different pump launching techniques is complex, since it involves many aspects: not only the transfer efficiency, but also the loss of brightness, the ease of manufacturing, the flexibility of handling, possible back-reflections, leakage of light from the fibre core back to the pump source, the option to maintain the polarization, etc.

LASERS AND AMPLIFIERS

The conceptually simplest option for generating high-power laser light is to build a fibre laser directly with some kind of mirrors on both ends. However, high-power fibre devices are very often built as laser-amplifier combinations, i.e., with a MOFA (*master oscillator fibre amplifier*) architecture. This concept has several advantages. It is easier to control the emission properties of a lower-power seed laser in terms of linewidth, laser noise, wavelength tunability, pulse generation, etc.

Also, the fibre of an amplifier has to stand only a power about equal to the output power, whereas in a laser the intracavity power is higher (even though fibre lasers as high-gain devices allow for strong output coupling).

Furthermore, it can be advantageous to use a modular design approach, where amplifier stages can be combined as required. In many cases, particularly when a low-power seed laser is used, one even uses several amplifier stages, typically with increasing mode areas and pump powers along the chain.

NANOSECOND PULSES

Many high-power lasers for material processing are Q-switched lasers, generating intense nanosecond pulses. In this regime, fibre devices are limited mainly in terms of peak power: apart from various nonlinearities, self-focusing introduces a hard limit at a few megawatts of peak power, which notably cannot be increased by using larger mode areas.

Actually achieved peak powers are usually below 1 MW, even for large mode area fibers. This means that, e.g., pulse energies of a few millijoules can not be combined with pulse durations of a few nanoseconds only.

On the other hand, the high gain of fibre devices allows the realization of very flexible MOPA devices with a gain-switched laser diode as seed laser. This approach, unlike Q switching of a laser, allows one, e.g., to modify the pulse duration independently of the pulse repetition rate.

ULTRASHORT PULSES

The large gain bandwidth of fibre amplifiers allows for the amplification of ultrashort pulses. However, particularly in this regime significant challenges arise from the strong fibre nonlinearity, because high-energy femtosecond pulses have huge peak powers.

In addition to the risk of fibre damage, strong nonlinear distortions can result from the fibre nonlinearity, and the high level of chromatic dispersion (including higher-order dispersion) can also be problematic. Particularly if pulse quality matters, these matters need to be carefully considered.

One possible approach is to make fibre-based chirped-pulse amplification systems, where the pulse duration within the amplifier is strongly increased, so that the effect of nonlinearities is accordingly decreased. An alternative to this approach is the amplification of parabolic pulses, where up-chirped pulses experiencing the gain and nonlinearity of the amplifier fibre evolve in a self-similar fashion, and their close to linear chirp makes it possible to obtain a high pulse quality with dispersive compression.

Although the mentioned techniques have allowed for extraordinary performance of fibre-based high-power ultrashort pulse sources (at least for multi-megahertz pulse repetition rates and correspondingly moderate pulse energies), these achievements are usually based on setups which involve a lot of free-space optics and hence lose many of the advantages of fibre systems.

An interesting direction is the development of all-fibre ultrashort pulse sources, possibly allowing for a final free-space pulse compressor (e.g. a transmission grating), but eliminating the need to launch pulses from free space into a fibre.

Prospects for Further Improvements

Even though the progress in the development of high-power fibre devices has been tremendous in recent years, various kinds of more or less severe limitations are now encountered, which are expected to slow this progress:

- The optical intensities in high-power fibre devices have been enormously increased. Now they are often close to the damage threshold of the material. Therefore, increased mode areas (→*large mode area fibers*) are used, but the limits of this approach have also been more or less reached, at least if very high beam quality is required.
- The power dissipation per unit length has reached values of the order of 100 W/m, which causes significant heating for air-cooled fibers. However, water cooling will allow for significant further increases in power. A longer fibre with lower doping concentration makes the cooling easier, but increases the effect of nonlinearities.
- Fibre nonlinearities can be disturbing in various ways. Even in continuous-wave devices, the Raman gain can become so high (tens of decibels) that a significant part of the power is transferred to a longer-wavelength Stokes wave, which does not experience laser gain. Single-frequency operation is even much more limited due to stimulated Brillouin scattering, although a number of measures are known to mitigate this effect to some extent. In devices for ultrashort pulses from mode-locked lasers, self-phase modulation can cause strong spectral broadening. In addition, there can be other problems such as nonlinear polarization rotation.

Due to such limitations, high-power fibre devices can – contrary to common opinions – usually not be considered as power-scalable in a strict sense, at least not well beyond the already achieved performance level.

This has important consequences e.g. when fibre laser technology is compared with that of thin-disk lasers. Of course, even without true power scalability, a lot can be done to improve various kinds of high-power fibre devices. An important issue is the further development of advanced fibre designs, e.g. with very large mode areas and single-mode guidance, as often realized with photonic crystal fibers. Various kinds of fibre devices are also important, e.g. special pump couplers, tapers for connecting fibers with different mode sizes and special cooling arrangements. Once the limit of the power per fibre has been reached, beam combining will be a further option – fibre devices appear suitable for this technique.

Also for ultrashort pulse amplifier systems there are various interesting options to mitigate and partially even to exploit the fibre nonlinearity, e.g. for further spectral broadening and subsequent pulse compression.

Beam Quality

The beam quality of a laser beam can be defined in different ways, but is essentially a measure of how tightly a laser beam can be focused under certain conditions (e.g. with a limited beam divergence). The most common ways to quantify the beam quality are:

- Tthe beam parameter product (BPP), i.e., the product of beam radius at the beam waist with the far-field beam divergence angle
- Tthe M^2 factor, defined as the beam parameter product divided by the corresponding product for a diffraction-limited Gaussian beam with the same wavelength
- Tthe inverse M^2 factor, which is high (ideally 1) for beams with high beam quality

The best possible beam quality is achieved for a diffraction-limited Gaussian beam, having $M^2 = 1$. This value is closely approached by many lasers, in particular by solid-state bulk lasers operating on a single transverse mode (→ *single-mode operation*) and by fibre lasers based on single-mode fibers, also by some low-power laser diodes (particularly VCSELs). On the other hand, in particular some high-power lasers (e.g. solid-state bulk lasers and semiconductor lasers such as diode bars) can have a very large M^2 of more than 100 or even well above 1000.

In solid-state lasers, this is often a result of thermally induced wavefront distortions in the gain medium and/or a mismatch of effective mode area and pumped area in the laser crystal, whereas in high-power semiconductor lasers the poor beam quality results from operation with a highly multimode waveguide. In both cases, the poor beam quality is associated with the excitation of higher-order resonator modes.

In the focus of a diffraction-limited beam (i.e., at the location where the beam radius reaches its minimum), the optical wavefronts are flat. Any scrambling of the wavefronts, e.g. due to optical components with poor quality, spherical aberrations of lenses, thermal effects in a gain medium, diffraction at apertures, or by parasitic reflections, can spoil the beam quality.

For monochromatic beams, the beam quality could in principle be restored e.g. with a phase mask which exactly compensates the wavefront distortions, but this is usually difficult in practice, even in cases where the distortions are stationary. A more flexible approach is to use adaptive optics in combination with a wavefront sensor. It is possible to some extent to improve the beam quality of a laser beam with a nonresonant mode cleaner or a mode cleaner cavity. This, however, leads to some loss of optical power. The brightness of a laser is determined by its output power together with its beam quality.

Note that the term *beam quality* is sometimes used with a qualitative meaning which has little to do with the focusability. For some applications, it is vital to obtain a smooth beam intensity profile, e.g. of Gaussian shape, whereas the beam divergence is not of interest. The "quality" of a laser beam may then not be characterized e.g. with an M^2: one beam may have a relatively small M^2 value but a multi-peaked beam profile, whereas another beam may have a smooth beam shape but a high divergence and thus a large M^2 value.

MEASUREMENT OF BEAM QUALITY

According to ISO Standard 11146, the beam quality factor M^2 can be calculated with a fitting procedure, applied to the measured evolution of the beam radius along the propagation direction (the so-called *caustic*). For correct results, a number of rules have to be observed, e.g. concerning the exact definition of the beam radius and the placing of data points.

There are commercially available beam profilers which can automatically perform beam quality measurements within a few seconds.

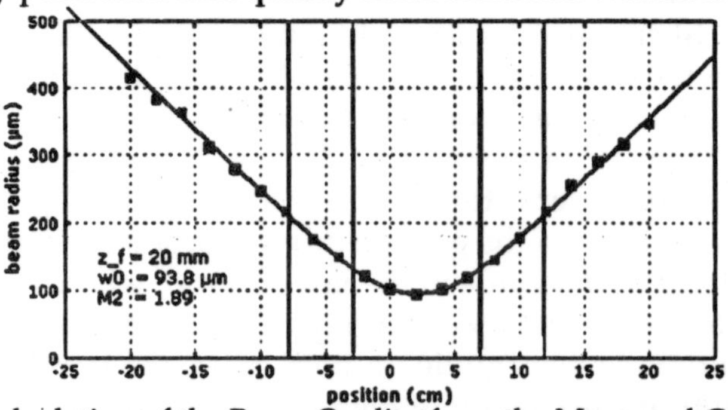

Fig. Calculation of the Beam Quality from the Measured Caustic.

They are normally based on the measurement of the beam profile at different positions. Beam profilers based on different measurement principles, e.g. CCD and CMOS cameras or rotating knife edges or slits, differ considerably in terms of the allowed ranges of beam radius and optical power, wavelength range, sensitivity to artifacts, etc.

For example, slit or knife-edge scanners can usually handle higher powers than cameras and can be precise for nearly Gaussian-shaped beams, whereas camera-based systems are usually more appropriate for complicated beam shapes. Other issues come into play for beams with temporally varying powers, e.g. for the output of Q-switched lasers. It may then be necessary to synchronize a shutter with the laser pulses.

Alternative measurement methods are based on the transmission through a mode-matched passive optical resonator or on wavefront sensors, e.g. Shack–Hartmann sensors. The full characterization of the laser beam then only requires analysis in a single plane.

Importance of Beam Quality for Applications

A high beam quality can be important e.g. when strong focusing of a beam is required. In the area of laser material processing, printing, marking, cutting and drilling require high beam qualities, whereas welding and various kinds of surface treatment are less critical in this respect, because they work with larger spots, so that direct application of high-power laser diodes with poor beam quality is possible.

For cutting and remote welding, a relatively high beam quality (with M^2 not much larger than 10) makes it possible to use a large working distance (i.e., a large distance between workpiece and focusing objective), which is highly desirable e.g. in order to protect the optics against debris and fumes.

Also, a high beam quality reduces the beam diameters in a beam delivery system, so that smaller and thus cheaper optical elements (e.g. mirrors and lenses) can be used. Furthermore, the increased effective Rayleigh length (for a given spot size) increases the tolerance for longitudinal alignment.

A large working distance, made possible by a high beam quality, is also important for the design of diode-pumped lasers when the pump beam has to go through various pieces of optics (e.g. a dichroic mirror) before reaching the laser crystal. A very high (close to diffraction-limited) beam quality, associated with a high spatial coherence, is often required for interferometers, optical data recording, laser microscopy, and the like.

Mode-locked lasers always have to have a high beam quality, since the excitation of higher-order transverse modes would disturb the pulse formation process.

Optimizing Laser Beam Quality

Crucial factors for obtaining a high beam quality from a solid-state bulk laser are:

- An optimized resonator design with suitable mode area (particularly in the gain medium) and low sensitivity to thermal lensing
- Good resonator alignment
- Minimized thermal effects, particularly from thermal lensing in the gain medium
- High-quality optical components (particularly concerning the gain medium)
- An optimized pump intensity distribution (sometimes requiring a pump source with good beam quality) – more easily achieved with end pumping than with side pumping.

Beam Quality in Nonlinear Optics

Beam quality is an issue not only for lasers, but also for nonlinear frequency conversion. While thermal lensing in nonlinear crystal materials occurs only at very high average power levels (because heating occurs only through weak parasitic absorption), the beam quality can be affected by other effects:

- Spatial walk-off can spatially shift the interacting beams, so that the overlap becomes weaker, and the interaction becomes spatially asymmetric.
- For strong conversion e.g. in a frequency doubler or an optical parametric amplifier, there can be strong depletion of the pump beam near the beam axis or even backconversion, in extreme cases leading to pronounced ring structures.

  Gain guiding can make such problems more severe. Beam quality issues have been shown to limit the power scalability of high-gain nonlinear frequency conversion devices.
- For ultrashort pulses, the group velocity mismatch and other effects can even lead to time-dependent beam quality.

Further, the use of a laser beam with poor beam quality in a nonlinear frequency conversion device can significantly spoil the conversion efficiency. Beam quality effects in nonlinear optics can be investigated with numerical computer models, which can simulate the evolution of the spatial (and possibly temporal) profiles of the involved beams.

RUBY LASER

A ruby laser is a solid-state laser that uses a synthetic ruby crystal as its gain medium. It was the first type of laser invented, and was first operated by Theodore H. "Ted" Maiman at Hughes Research Laboratories on 1960-05-16.The ruby laser produces pulses of visible light at a wavelength of 694.3 nm, which appears as deep red to human eyes.

Typical ruby laser pulse lengths are on the order of a millisecond. These short pulses of red light are visible to the human eye, if the viewer carefully watches the target area where the pulse will fire.

Solid-state lasers are lasers based on solid-state gain media such as crystals or glasses doped with rare earth or transition metal ions, or semiconductor lasers. (Although semiconductor lasers are of course also solid-state devices, they are often not included in the term solid-state lasers.)

Ion-doped solid-state lasers (also sometimes called *doped insulator lasers*) can be made in the form of bulk lasers, fibre lasers, or other types of waveguide lasers. Solid-state lasers may generate output powers between a few milliwatts and (in high-power versions) many kilowatts.

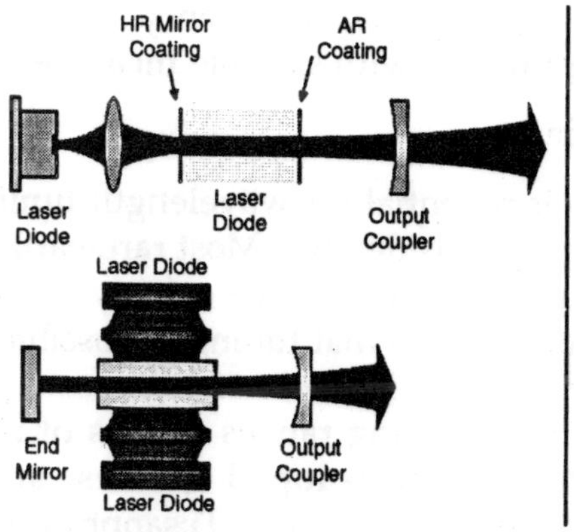

Fig. Typical Setups of Solid-State Bulk Lasers.

Optical Pumping and Energy Storage

Many solid-state lasers are optically pumped with flash lamps or arc lamps. Such pump sources are relatively cheap and can provide very high powers. However, they lead to a fairly low power efficiency, moderate lifetime, and strong thermal effects such as thermal lensing in the gain medium.

For such reasons, laser diodes are very often used for pumping solid-state lasers. Such *diode-pumped solid-state lasers* (DPSS lasers, also called *all-solid-state lasers*) have many advantages, in particular a compact setup, long lifetime, and often very good beam quality. Therefore, their share of the market is rapidly rising.

The laser transitions of rare-earth or transition-metal-doped crystals or glasses are normally weakly allowed transitions, i.e., transitions with very low oscillator strength, which leads to long radiative upper-state lifetimes and consequently to good energy storage, with upper-state lifetimes of microseconds to milliseconds. Although this is beneficial for nanosecond pulse generation, it can also lead to unwanted spiking phenomena in continuous-wave lasers, e.g. when the pump source is switched on.

Pulse Generation

The long upper-state lifetimes makes solid-state lasers very suitable for Q switching: the laser crystal can easily store an amount of energy which, when released in the form of a nanosecond pulse, leads to a peak power which is orders of magnitude above the achievable average power. Bulk lasers can thus easily achieve millijoule pulse energies and megawatt peak powers.

In mode-locked operation, solid-state lasers can generate ultrashort pulses with durations measured in picoseconds or femtoseconds. With passive mode locking, they have a tendency for Q-switching instabilities, if these are not suppressed with suitable measures.

Wavelength Tuning

In terms of their potential for wavelength tuning, different types of solid-state lasers differ considerably. Most rare-earth-doped laser crystals, such as Nd:YAG and Nd:YVO_4, have a fairly small gain bandwidth of the order of 1 nm or less, so that tuning is possible only within a rather limited range.

On the other hand, tuning ranges of tens of nanometers and more are possible with rare-earth-doped glasses, and particularly with transition-metal-doped crystals such as Ti:sapphire, Cr:LiSAF and Cr:ZnSe (→ *vibronic lasers*).

Types of Solid-State Lasers

Examples of different types of solid-state lasers are:

- Small diode-pumped Nd:YAG (→*YAG lasers*) or Nd:YVO_4 lasers (→ *vanadate lasers*) often operate with output powers between a few milliwatts (for miniature setups) and a few watts. Q-switched versions generate pulses with durations of a few nanoseconds, microjoule pulse energies and peak powers of many kilowatts. Intracavity frequency doubling can be used for green output.
- Single-frequency operation, typically achieved with unidirectional ring lasers (e.g. NPROs = nonplanar ring oscillators) or microchip lasers, allows for operation with very small linewidth in the lower kilohertz region.
- Larger lasers in side-pumped or end-pumped configurations, having the geometry of rod lasers, slab lasers or thin-disk lasers,

are suitable for output powers up to several kilowatts. Particularly thin-disk lasers can still offer very high beam quality, and also a high power efficiency.

- Q-switched Nd:YAG lasers are still widely used in lamp-pumped versions. Pulsed pumping allows for high pulse energies, whereas the average output powers are often moderate (e.g. a few watts). The cost of such lamp-pumped lasers is lower than for diode-pumped versions with similar output powers.
- Fibre lasers are a special kind of solid-state lasers, with a high potential for high average output power, high power efficiency, high beam quality, and broad wavelength tunability.

Applications

Ruby lasers have declined in use with the discovery of better lasing media. They are still used in a number of applications where short pulses of red light are required. Holographers around the world produce holographic portraits with ruby lasers, in sizes up to a metre squared.

The red 694 nm laser light is preferred to the 532 nm green light of frequency-doubled Nd:YAG..Many non-destructive testing labs use ruby lasers to create holograms of large objects such as aircraft tires to look for weaknesses in the lining. Ruby lasers were used extensively in tattoo and hair removal, but are being replaced by alexandrite lasers and Nd:YAG lasers in this application.

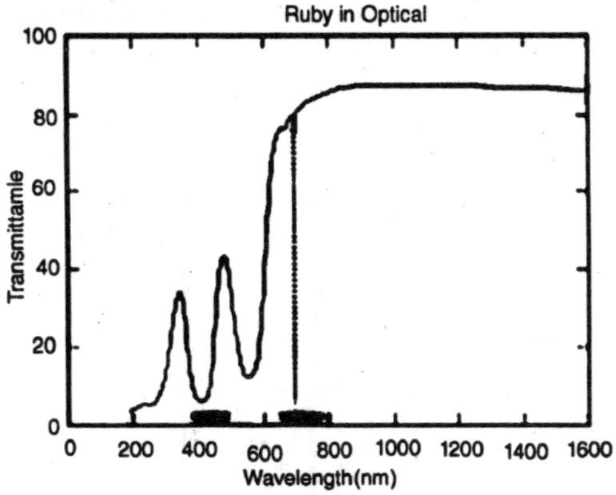

Fig. Transmittance of Ruby in Optical and Near-IR Spectra.

Design

The ruby laser is a three level solid state laser. The active laser medium (laser gain/amplification medium) is a synthetic ruby rod that is energized through optical pumping, typically by a xenon flash lamp. In early examples, the rod's ends had to be polished with great precision, such that the ends of the rod were flat to within a quarter of a wavelength of the output light, and parallel to each other within a few seconds of arc. The finely polished ends of the rod were silvered: one end completely, the other only partially. The rod with its reflective ends then acts as a Fabry-Pérot etalon (or a Gires-Tournois etalon). Modern lasers often use rods with ends cut and polished at Brewster's angle instead. This eliminates the reflections from the ends of the rod; external dielectric mirrors then are used to form the optical cavity. Curved mirrors are typically used to relax the alignment tolerances

Coherence

Coherence is one of the most important concepts in optics and is strongly related to the ability of light to exhibit interference effects. A light field is called *coherent* when there is a fixed phase relationship between the electric field values at different locations or at different times. *Partial coherence* means that there is some (although not perfect) correlation between phase values.

Spatial Versus Temporal Coherence

There are two very different aspects of coherence:

- *Spatial coherence* means a strong correlation (fixed phase relationship) between the electric fields at different locations across the beam profile. For example, within a cross-section of a beam from a laser with diffraction-limited beam quality, the electric fields at different positions oscillate in a totally correlated way, even if the temporal structure is complicated by a superposition of different frequency components. Spatial coherence is the essential prerequisite of the strong directionality of laser beams.

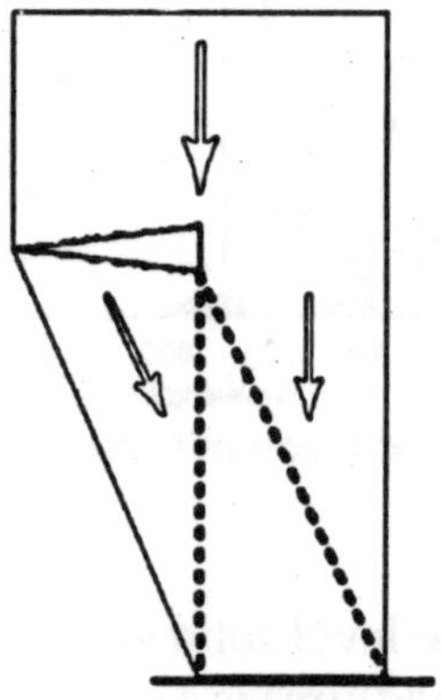

Fig. A Prism is Inserted into a Spatially Coherent Laser Beam, Generating an Interference Pattern on the Screen.

- *Temporal coherence* means a strong correlation between the electric fields at one location but different times. For example, the output of a single-frequency laser can exhibit a very high temporal coherence, as the electric field temporally evolves in a highly predictable fashion: it exhibits a clean sinusoidal oscillation over extended periods of time.

Lasers have the potential for generating beams (e.g. Gaussian beams) with very high spatial coherence, and this is perhaps the most fundamental difference between laser light and radiation from other light sources. High spatial coherence arises from the existence of resonator modes, which define spatially correlated field patterns. In situations where only a single resonator mode has sufficient laser gain to oscillate, a single longitudinal mode can be selected, obtaining single-frequency operation with very high *temporal* coherence as well. The difference between spatial and temporal coherence.

Fig. Electric Field Distribution Around the Focus of a Gaussian Laser Beam with Perfect Spatial and Temporal Coherence.

A beam with high spatial coherence, but poor temporal coherence. The wavefronts are formed as above, and the beam quality is still very high, but the amplitude and phase of the beam varies along the propagation direction. Note that both the local amplitude and the spacing of the wavefronts vary to some extent. Such a beam can be generated e.g. from the output of a supercontinuum source.

Fig. A Laser Beam with High Spatial Coherence, but Poor Temporal Coherence.

A laser beam with poor spatial coherence, but high temporal coherence. The wavefronts are deformed, and this results in a high beam divergence and poor beam quality. On the other hand, the beam is monochromatic, so that the spacing of the deformed wavefronts remains constant. Such a beam can result from a single-frequency laser, when its output is sent through some optically inhomogeneous material.

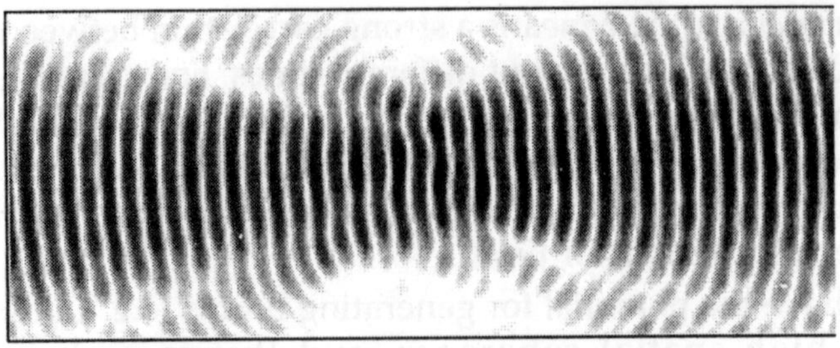

Fig. A Laser Beam with Poor Spatial Coherence, but High Temporal Coherence.

Quantifying Coherence

There are different ways to quantify the degree of coherence:

- Correlation functions specify the degree of correlation as a function of a spatial or temporal distance.
- Fringe visibility parameters essentially specify the visibility (contrast) of an interference pattern generated by superposition of two electric fields.
- The coherence time quantifies the degree of temporal coherence via the time over which coherence is lost.
- The coherence length is the coherence time times the vacuum velocity of light, and thus also characterizes the temporal (not spatial!) coherence via the propagation length (and thus propagation time) over which coherence is lost.
- The linewidth of a single-frequency laser is also strongly related to temporal coherence: a narrow linewidth (high monochromaticity) means high temporal coherence.

The relationship between optical bandwidth and temporal coherence can be non-trivial. For example, a pulse train from a mode-locked laser can have a broad overall bandwidth, with the Fourier spectrum consisting of discrete very narrow lines (→ *frequency combs*).

The temporal coherence can be very high in the sense that there are strong field correlations for large time delays which are close to integer multiples of the pulse period.

Importance of Coherence in Applications

Some applications need light with very high spatial and temporal coherence. This applies, e.g., to many variations of interferometry, holography, and some types of optical sensors (e.g. fibre-optic sensors). Such features are also important for the technique of coherent beam combining.

For other applications, the coherence of the light used should be as *low* as possible.

For example, very low temporal coherence (but combined with high spatial coherence) is required for *coherence tomography*, where images are created with a kind of interferometry, and a high spatial resolution requires low temporal coherence.

Suitable light sources for such applications can be based on amplified spontaneous emission (ASE) from a laser amplifier (→ *superluminescent sources*) or on supercontinuum generation in nonlinear media.

A low degree of temporal coherence can also be beneficial for laser projection displays, imaging and pointer applications, as it reduces the tendency for speckle and similar interference effects.

COHERENCE IN QUANTUM OPTICS

In quantum optics, the term *coherence* is often used for the state of light-emitting atoms or ions. In that case, coherence refers to a phase relationship between the complex amplitudes corresponding to electronic states.

This is important, e.g., in the context of lasing without inversion. There is also the term "coherent states" of the light field, which has yet another meaning.

Coherence Time

The coherence time can be used for quantifying the degree of temporal coherence of light. In coherence theory, it is essentially defined as the time over which the field correlation function decays.

This correlation (or coherence) function is:

$$g(\tau) = \frac{\left\langle E^x(t)E(t+\tau)\right\rangle}{\left\langle E^x(t)E(t)\right\rangle}$$

where *E(t)* is the complex electric field at a certain location. This function is 1 for = 0 and usually decays monotonically for larger time delays.

For an arbitrary shape of this function, the coherence time can be defined by:

$$\tau_{coh} = \int_{-\infty}^{+\infty} |g(\tau)|^2 \, d\tau.$$

In the case of an exponential coherence decay (as it occurs e.g. for a laser with quantum noise influences only), this is the same as the exponential decay time.

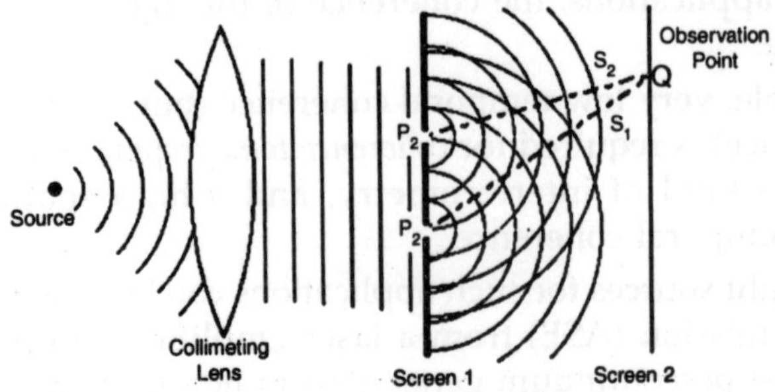

Fig. Temporal Coherence.

Knowledge of the coherence time (i.e., a single number) can be useful when the shape of the coherence function (or the shape of the Fourier spectrum) is approximately known. Obviously, however, the specification of the coherence time (or the linewidth) alone does not constitute a full characterization of the coherence.

The coherence time is intimately linked with the linewidth of the radiation, i.e., the width of its spectrum. In the case of an exponential coherence decay as above, the spectrum has a Lorentzian shape, and the (full width at half-maximum) linewidth is

$$\Delta\nu = \frac{1}{\pi\tau_{coh}}.$$

The constant factor in this equation is in general different for other shapes of the coherence function (e.g. roughly twice as high for a Gaussian shape). Conversely, the linewidth can be used for estimating the coherence time, but the conversion depends on the spectral shape. In cases where the frequency noise spectrum is not flat but rises strongly at small noise frequencies, there can be a significant degree of coherence even for time delays well above the inverse linewidth; this issue is important, e.g., in the context of self-heterodyne linewidth measurement.

Instead of the coherence time, it is common to specify the coherence length, which is simply the coherence time times the vacuum velocity of light, and thus also quantifies temporal (rather than spatial) coherence. Lasers, particularly single-frequency solid-state lasers, can have long coherence times, compared with the duration of an optical cycle; values of several milliseconds are possible.

COHERENCE LENGTH

The coherence length can be used for quantifying the degree of temporal (not spatial!) coherence as the propagation length (and thus propagation time) over which coherence degrades significantly. It is defined as the coherence time times the vacuum velocity of light.

For light with a Lorentzian optical spectrum, the coherence length can be calculated as:

$$L_{coh} = C\tau_{coh} = \frac{C}{\pi\Delta\nu}$$

where $\Delta\nu$ is the (full width at half-maximum) linewidth (optical bandwidth).

However, such relations are not valid in cases where the coherence function has a more complicated shape, as is the case for, e.g., a frequency comb.

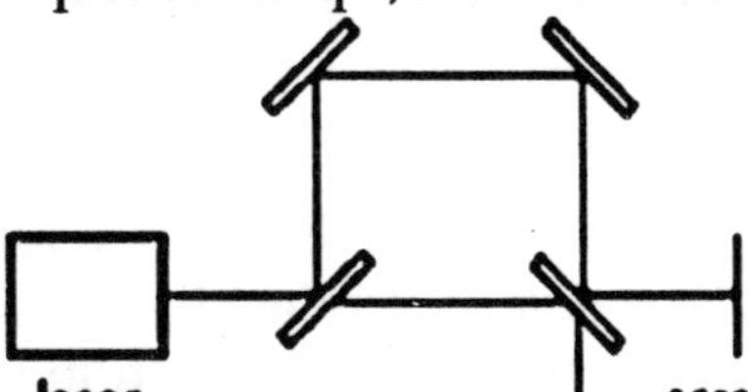

Fig. Setup of an Interferometer

The reason for often using the term *coherence length* instead of *coherence time* is that the optical time delays involved in some experiment are often determined by optical path lengths. The interferometer shows pronounced interference fringes only if the coherence length of the laser light is at least as long as the path-length difference of the two arms.

Also, in a setup for making holographic recordings, coherence between two beams with a somewhat different optical path length is required, so that the coherence length of the light source should be longer than the maximum occurring path-length difference.

In addition to holography, a number of other applications may require a certain coherence length. Lasers, particularly single-frequency solid-state lasers, can have very long coherence lengths, e.g. 9.5km for a Lorentzian spectrum with a linewidth of 10kHz. The coherence length is limited by phase noise which can result from, e.g., spontaneous emission in the gain medium.

COHERENCE LENGTH IN NONLINEAR OPTICS

An unfortunate use of the term *coherence length* is common in nonlinear optics: for example, in second-harmonic generation, the coherence length is often understood as the length over which fundamental and harmonic wave get out of phase (more precisely, the phase difference accumulated over this length is δ).

This is inconsistent with the general notion of coherence, because a predictable phase relationship (strong phase correlation) is definitely maintained over more than this length, although there is a *systematic* evolution of the relative phase.

Chapter 11

Mass and Energy

The fact that feeding energy into a body increases its mass suggests that the mass m_0 of a body at rest, multiplied by c^2, can be considered as a quantity of energy. The truth of this is best seen in interactions between elementary particles.

For example, there is a particle called a *positron* which is exactly like an electron except that it has positive charge. If a positron and an electron collide at low speed (so there is very little kinetic energy) they both disappear in a flash of electromagnetic radiation. This can be detected and its energy measured. It turns out to be $2m_0c^2$ where m_0 is the mass of the electron. Thus particles can "vapourize" into pure energy, that is, electromagnetic radiation. The energy m_0c^2 of a particle at rest is called its "rest energy". Note, however, that an electron can only be vapourized by meeting with a positron, and there are very few positrons around normally, for obvious reasons-they just don't get far.

Einstein's Box

An amusing "experiment" on the equivalence of mass and energy is the following: consider a closed box with a flashlight at one end and light-absorbing material at the other end. Imagine the box to be far out in space away from gravitational fields or any disturbances. Suppose the light flashes once, the flash travels down the box and is absorbed at the other end.

Now it is known from Maxwell's theory of electromagnetic waves that a flash of light carrying energy E also carries momentum $p = E/c$. Thus, as the flash leaves the bulb and goes down the tube, the box recoils, like a gun, to conserve overall momentum. Suppose the whole apparatus has mass M and recoils at velocity v. Of course, $v \ll c$.

Then from conservation of momentum:

$Mv = E/c$.

After a time $t = L/C$ the light hits the far end of the tube, is absorbed, and the whole thing comes to rest again.

How far did the box move?

It moved at speed v for time t, so it moved distance $vt = vL/c$.

From the conservation of momentum equation above, we see that $v = E/Mc$, so the distance d the box moved over before stopping is given by:

$$d = \frac{EL}{Mc^2}$$

Now, the important thing is that there are *no* external forces acting on this system, so *the centre of mass cannot have moved!*

The only way this makes sense is to say that to counterbalance the mass M moving d backwards, the light energy *must have transferred a small mass m,* say, the length L of the tube so that

$$Md = mL$$

and balance is maintained. From our formula for d above, we can figure out the necessary value of m,

$$m = \frac{M}{L}d = \frac{M}{L}\cdot\frac{EL}{Mc^2} = \frac{E}{c^2}$$

so

$$E = mc^2.$$

We have therefore established that *transfer of energy implies transfer of the equivalent mass.* All we had to assume was that the centre of mass of an isolated system, initially at rest, remains at test if no external forces act, and that electromagnetic radiation carries momentum E/c, as predicted by Maxwell's equations and experimentally established.

But how is this mass transfer physically realised? Is the front end of the tube really heavier after it absorbs the light? The answer is yes, because it's a bit hotter, which means its atoms are vibrating slightly faster — and faster moving objects have higher mass.

Mass and Potential Energy

Suppose now at the far end of the tube we have a hydrogen atom at rest. As we shall discuss later, this can be thought of as a proton with an electron bound to it by electrostatic attraction, and it is known that a flash of light having total energy 13.6eV is just enough to tear the electron away, so in the end the proton and electron are at rest far away from each other.

The energy of the light was used up dragging the proton and electron apart — that is, it went into potential energy. Yet the light *is* absorbed by this process, so from our argument above the right hand end of the tube *must become heavier.*

That is to say, *a proton at rest plus a (distant) electron at rest weigh more than a hydrogen atom* by E/c^2, with E equal to 13.6eV. Thus, Einstein's box forces us to conclude that increased *potential* energy in a system also entails the appropriate increase in mass.

It is interesting to consider the hydrogen atom dissociation in reverse — if a slow moving electron encounters an isolated proton, they may combine to form a hydrogen atom, emitting 13.6eV of electromagnetic radiation energy as they do so. Clearly, then, the hydrogen atom remaining has that much less energy than the initial proton + electron.

The actual mass difference for hydrogen atoms is about one part in 10^8. This is typical of the energy radiated away in a violent chemical reaction — in fact, since most atoms are an order of magnitude or more heavier than hydrogen, a part in 10^9 or 10^{10} is more usual. However, things are very different in nuclear physics, where the forces are stronger so the binding is tighter.

We shall discuss this later, but briefly mention an example-a hydrogen nucleus can combine with a lithium nucleus to give two helium nuclei, and the mass shed is 1/500 of the original. This reaction has been observed, and all the masses involved are measurable. The actual energy emitted is 17 MeV. This is the type of reaction that occurs in hydrogen bombs. Notice that the energy released is at least a million times more than the most violent chemical reaction.

Final example — give a ballpark estimate of the change in mass of a million tons of TNT on exploding. The TNT molecule is about a hundred times heavier than the hydrogen atom, and gives off a few eV on burning. So the change in weight is of order $10^{-10} \times 10^6$ tons, about a hundred grams. In a hydrogen bomb, this same mass to energy conversion would take about fifty kilograms of fuel.

Einstein's Box is a Fake

Although Einstein's box argument is easy to understand, and gives the correct result, it is based on a physical fiction — the rigid box. If we had a rigid box, or even a rigid stick, all our clock synchronization problems would be over — we could start clocks at the two ends of the stick simultaneously by nudging the stick from one end, and, since it's a rigid stick, the other end would move instantaneously. Actually there are no such materials.

All materials are held together by electromagnetic forces, and pushing one end causes a wave of compression to travel down the stick. The electrical forces between atoms adjust at the speed of light, but the overall wave travels far more slowly because each atom in the chain must accelerate for a while before it moves sufficiently to affect the next one measurably.

So the light pulse will reach the other end of the box before it has begun to move! Nevertheless, the wobbling elastic box *does* have a net recoil momentum, which it does lose when the light hits the far end. So the basic point is still valid. French gives a legitimate derivation by replacing the box by its two (disconnected) ends, which he treats as a single system.

Energy and Momentum in Lorentz Transformations

We have a formula for the total energy E = K.E. + rest energy,

$$E = mc^2 = \frac{m0c^2}{\sqrt{1 - v^2/c^2}}$$

so we can see how total energy varies with speed.

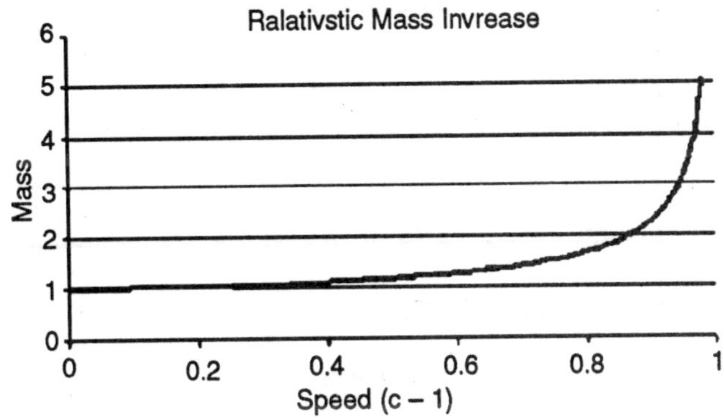

The momentum varies with speed as

$$P = mv = \frac{m_0 v}{\sqrt{1 - v^2/c^2}}$$

How Does the Total Energy of a Particle Depend on Momentum?
It turns out to be useful to have a formula for E in terms of p.
Now,

$$E^2 = m^2c^4 = \frac{m_0^2c^4}{1 - v^2/c^2}$$

so,

$$m^2c^4\left(1 - v^2/c^2\right) = m_0^2c^4$$

$$m^2c^4 - m^2v^2c^2 = m_0^2c^4$$

$$m^2c^4 = E^2 = m_0^2c^4 + m^2c^2v^2$$

hence using $p = mv$ we find:

$$E = \sqrt{m_0^2c^4 = c^2p^2}$$

If p is very small, this gives

$$E \approx m_0c^2 + \frac{p^2}{2m_0},$$

the usual classical formula.

If p is very large, so $c^2p^2 >> m_0{}^2c^4$, the approximate formula is $E = cp$.

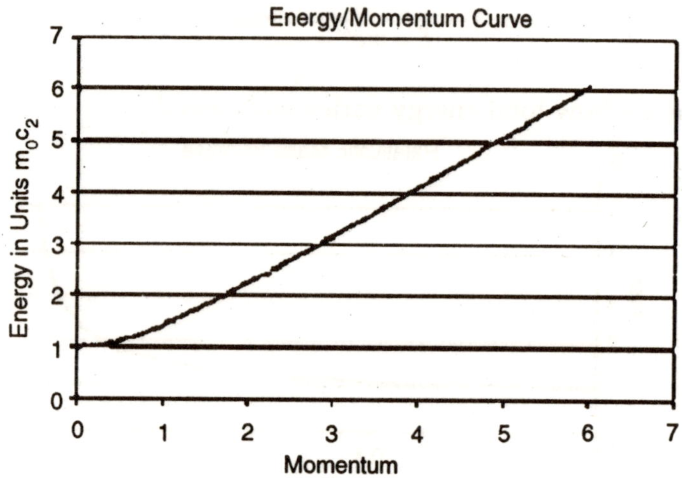

The High Kinetic Energy Limit

Notice that this high energy limit is just the energy-momentum relationship Maxwell found to be true for light, for all p. This could only be true for *all* p if $m_0{}^2c^4 = 0$, that is, $m_0 = 0$.

Light is in fact composed of "photons"—particles having zero "rest mass", as we shall discuss later.

The "rest mass" of a photon is meaningless, since they're never at rest—the energy of a photon

$$E = mc^2 = \frac{m_0c^2}{\sqrt{1 - v^2 / c^2}}$$

is of the form 0/0, since $m_0 = 0$ and $v = c$, so "m" can still be nonzero. That is to say, the mass of a photon is really all K.E. mass.

For very fast electrons, such as those produced in high energy accelerators, the additional K.E. mass can be thousands of times the rest mass.

For these particles, we can neglect the rest mass and take $E = cp$.

Transforming Energy and Momentum to a New Frame

$$\vec{P} = m\vec{v} = \frac{m_0\vec{v}}{\sqrt{1-v^2/c^2}}$$

$$E = mc^2 = \sqrt{m_c^2} = \sqrt{m_0^2c^4 + c^2\vec{p}^2}$$

Notice we can write this last equation in the form:

$$E^2 - c^2\vec{p}^2 = m_0^2c^4$$

That is to say, $E^2 - c^2\vec{p}^2$ depends *only* on the rest mass of the particle and the speed of light. It does not depend on the velocity of the particle, so it must be the same—for a particular particle—in all inertial frames.

This is reminiscent of the invariance of $\vec{x}^2 - c^2t^2$, the interval squared between two events, under the Lorentz transformations. One might guess from this that the laws governing the transformation from E, p in one Lorentz frame to E', p' in another are similar to those for x,t. We can actually derive the laws for E, p to check this out.

As usual, we consider all velocities to be parallel to the x-axis.

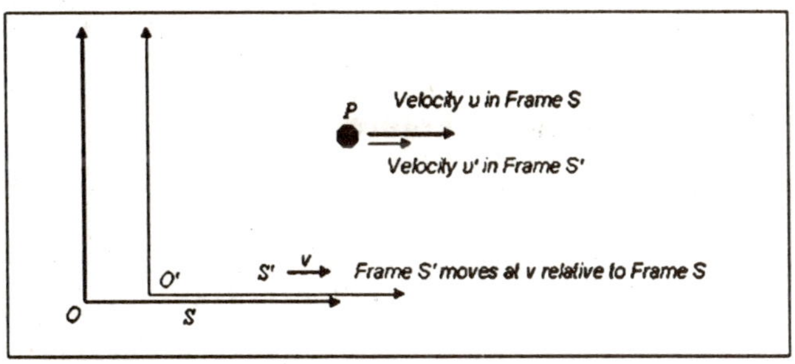

We take the frame S' to be moving in the x-direction at speed v relative to S.

Consider a particle of mass m_0 (rest mass) moving at u' in the x' direction in frame S', and hence at u along x in S, where

$$u = \frac{u' + v}{1 + vu'/c^2}$$

The energy and momentum in S' are

$$E = \frac{m_0c^2}{\sqrt{1-u'^2/c^2}} \cdot p' = \frac{m_0u'}{\sqrt{1-u'^2/c^2}}$$

and in S:

$$E = \frac{m_0c^2}{\sqrt{1-u'^2/c^2}} \cdot p = \frac{m_0u}{\sqrt{1-u'^2/c^2}}$$

Thus,

$$E = \frac{m_0 c^2}{\sqrt{1 - \left(\frac{u' + v}{1 + vu'/c^2}\right)}}$$

giving,

$$E = \frac{m_0 c^2 \left(1 + vu'/c^2\right)}{\sqrt{\left(1 - v^2/c^2\right)\left(1 - u^2/c^2\right)}}$$

from which it is easy to show that

$$E = \frac{1}{\sqrt{1 - v^2/c^2}}(E' + vp')$$

Similarly, we can show that

$$P = \frac{P' + vE'/c^2}{\sqrt{1 - v^2/c^2}}$$

These are the Lorentz transformations for energy and momentum of a particle—it is easy to check that

$$E^2 - c^2 p^2 = E'^2 - c^2 p'^2 = m_0^2 c^4$$

Photon Energies in Different Frames

For a zero rest mass particle, such as a photon, $E = cp$, $E^2 - c^2p^2 = 0$ in all frames.

Thus,

$$E = \frac{Evp'}{\sqrt{1 - v^2/c^2}} = \frac{E + vE'/c}{\sqrt{1 - v^2/c^2}} = E\sqrt{\frac{1 + v/c}{1 - v/c}}.$$

Since $E = cp$, $E' = cp'$ we also have

$$P = p'\sqrt{\frac{1 + v/c}{1 - v/c}}.$$

Notice that the ratios of photon *energies* in the two frames coincides with the ratio of photon *frequencies* found in the Doppler shift. As we shall see when we cover quantum mechanics, the photon energy is proportional to the frequency, so these two must of course transform in identical fashion.

But it's interesting to see it come about this way. Needless to say, relativity gives us no clue on what the constant of proportionality (Planck's constant) is: it must be measured experimentally. But the *same* constant plays a role in all quantum phenomena, not just those concerned with photons.

Transforming Energy into Mass

Pion Production

We have mentioned how, using a synchrocyclotron, it is possible to accelerate protons to relativistic speeds. The rest energy of a proton $m_p c^2$ is 938 MeV, using here the standard high energy physics energy unit: 1 MeV = 10^6 eV. The neutron is a bit heavier—$m_n c^2$ = 940 MeV. (The electron is 0.51 MeV). Thus to accelerate a proton to relativistic speeds implies giving it a K.E. of order 1,000 MeV, or 1 GeV.

The standard operating procedure of high energy physicists is to accelerate particles to relativistic speeds, then smash them into other particles to see what happens. For example, fast protons will be aimed at protons at rest (hydrogen atoms, in other words—the electron can be neglected). These proton-proton collisions take place inside some kind of detection apparatus, so the results can be observed. One widely-used detector is the bubble chamber: a transparent container filled with a superheated liquid.

The electric field of a rapidly moving charged particle passing close to a molecule can dislodge an electron, so an energetic particle moving through the liquid leaves a trail of ionized molecules. These give centres about which bubbles can nucleate. The bubbles grow rapidly and provide a visible record of the particle's path.

What is actually observed in *p-p* scattering at relativistic energies is that often more particles come out than went in—particles called pions, π^+, π^0, π^- can be created. The π^0 is electrically neutral, the π^+ has exactly the same amount of charge as the proton. It is found experimentally that total electric charge is always conserved in collisions, no matter how many new particles are spawned, and total baryon number (protons + neutrons) is conserved.

Possible scenarios include:

$$p + p \rightarrow p + p\pi^0$$

and

$$p + p \rightarrow p + n + \pi^*$$

The neutral pion mass is 135 MeV, the charged pions have mass 140 MeV, where we follow standard high energy practice in calling mc^2 the "mass", since this is the energy equivalent, and hence the energy which, on creation of the particle in a collision, is taken from kinetic energy and stored in mass.

However, an incoming proton with 135 MeV of kinetic energy will not be able to create a neutral pion in a collision with a stationary proton. This is because the incoming proton also has momentum, and the collision conserves momentum, so some of the particles after the collision must have momentum and hence kinetic energy.

The simplest way to figure out just how much energy the incoming proton needs to create a neutral pion is to go to the centre of mass frame, where initially two protons are moving towards each other with equal and opposite velocities, there is no total momentum. Obviously, in this frame the least possible K.E. must be just enough to create the π^0 with all the final state particles (p,p,π^0) at rest. Thus if the relativistic mass of the incoming protons in the centre of mass frame is m, the total energy

$$E = 2mc^2 = 2m_pc^2 + m_pc^2$$

Putting in the proton and pion masses from above, and using $m = m_p / \sqrt{1-(v^2/c^2)}$, we find the two incoming protons must both be traveling at 0.36*c*.

Recall that this is the speed in the centre of mass frame, and for practical purposes, like designing the accelerator, we need to know the energy necessary in the "lab" frame—that in which one of the protons is initially at rest. The two frames obviously have a relative speed of 0.36*c*, so to get the speed of the incoming proton in the lab frame we must add a velocity of 0.36*c* to one of 0.36*c* using the relativistic addition of velocities formula, which gives 0.64*c*. This implies the incoming proton has a relativistic mass of 1.3 times its rest mass, and thus a K.E. around 280 MeV.

Thus to create a pion of rest energy 135 MeV, it is necessary to give the incoming proton at least 290 MeV of kinetic energy. This is called the "threshold energy" for pion production. The "inefficiency" arises because momentum is also conserved, so there is still considerable K.E. in the final particles.

ANTIPROTON PRODUCTION

On raising the energy of the incoming proton further, more particles are produced, including the "antiproton"—a negatively charged heavy particle which will annihilate a proton in a flash of energy. It turns out experimentally that an antiproton can only be produced accompanied by a newly created proton,

$$p + p \rightarrow p + p + p + \overline{p}$$

Notice we could have conserved electric charge with less energy with the reaction

$$p + p \rightarrow p + p + \pi^* + \overline{p}$$

but this doesn't happen—so charge conservation isn't the only constraint on which particles can be produced. In fact, what we are seeing here is experimental confirmation that the conservation of baryon number, which at the low energies previously discussed in the context of pion production just meant that the total number of protons plus neutrons stayed fixed, is generalized at high energies to include antiparticles with negative baryon number, -1 for the antiproton.

Thus baryon number conservation becomes parallel to electric charge conservation, new particles can always be produced at high enough energies provided the total new charge and the total new baryon number are both zero. We should emphasize again that these are *experimental* results gathered from examining millions of collisions between relativistic particles.

One of the first modern accelerators, built at Berkeley in the fifties, was designed specifically to produce the antiproton, so it was very important to calculate the antiproton production threshold correctly! This can be done by the same method we used above for pion production, but we use a different trick here which is often useful. We have shown that on transforming the energy and momentum of a particle from one frame to another.

$$E^2 - c^2p^2 = E'^2 - c^2p'^2$$

Since the Lorentz equations are linear, if we have a system of particles with total energy E and total momentum p in one frame, E', p' in another, it must again be true that.

$$E^2 - c^2\vec{p}^2 = E^2 - c^2\vec{p}'^2$$

We can use this invariance to get lab frame information from the centre of mass frame. Noting that in the centre of mass (CM) frame the momentum is zero, and in the lab frame the momentum is all in the incoming proton,

$$E_m^2 - \left((\mathrm{m_{in}} + m_0)c^2\right)^2 - c^2p_{in}^2$$

where here m_0 is the proton rest mass, m_{in} is the relativistic mass of the incoming proton. At the antiproton production threshold, $E_{cm} = 4m_0c^2$, so

$$16m_0^2c^4 = m_{in}^2 - c^2p_{in}^2c^4 + 2\,\mathrm{m_{in}}\,c^2m_0c^2 + m_0^2c^4 - c^2p_{in}^2$$

and using,

$$m_{in}^2c^4 - c^2p_{in}^2 = m_0^2c^4$$

we find,

$$2\left(m_{in}c^2\right)\left(m_0c^2\right) + 2\left(m_0c^2\right)^2 = 16\left(m_0c^2\right)^2,$$

so,

$$m_{in}c^2 = 7m_0c^2.$$

Therefore to create two extra particles, with total rest energy $2m_0c^2$, it is necessary for the incoming proton to have a kinetic energy of $6m_0c^2$. The Berkeley Gevatron had design energy 6.2 GeV.

As we go to higher energies, this "inefficiency" gets worse—consider energies such that the kinetic energy >> rest energy, and assume the incoming particle and the target particle have the same rest mass, m_0, with the incoming particle having relativistic mass m.

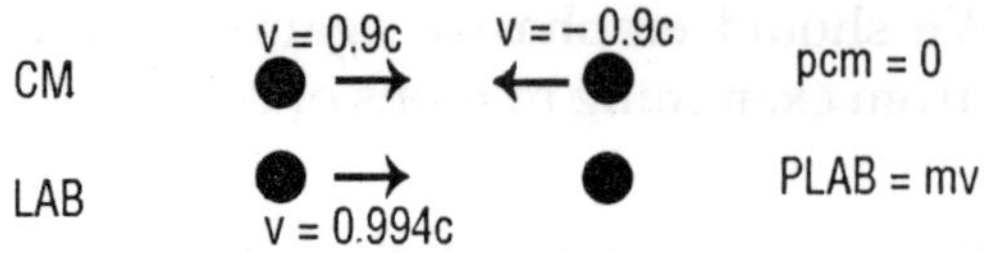

Comparing the centre of mass energy with the lab energy at these high energies,

$$
\begin{aligned}
E_{LAB} &= (m + m_0)c^2, \\
E^2{}_{CM} &= E_{LAB} - p^2_{LAB}c^2, \\
&= m^2c^4 + 2mc^2m_0c^2 + m_0^2c^4 - P^2_{LAB}c^2 \\
&= 2m_0c^2\left(mc^2 + m_0c^2\right)
\end{aligned}
$$

For $m >> m_0$,

$$E^2_{CM} \approx 2m_0c^2mc^2 \approx 2m_0c^2.E_{LAB}$$

so

$$Ecm \approx \sqrt{2m_0c^2E_{LAB}},$$

ultimately one must *quadruple* the lab energy to double the centre of mass energy.

Chapter 12

Relativity

RELATIVITY CONNECTS ELECTRIC AND MAGNETIC FIELDS

Suppose we have an infinitely long straight wire, having a charge density of electrons of $-\lambda$ coulombs per meter, all moving at speed v to the right (recall typical speeds are centimeters per minute) and a neutralizing fixed background of positive charge, also of course λ coulombs per meter. The current in the wire has magnitude $I = \lambda v$ (and actually is flowing to the left, since the moving electrons carry negative charge).

Suppose also that a positive charge q is outside the wire, a distance r from the axis, and this outside charge is moving at the same exact velocity as the electrons in the wire.

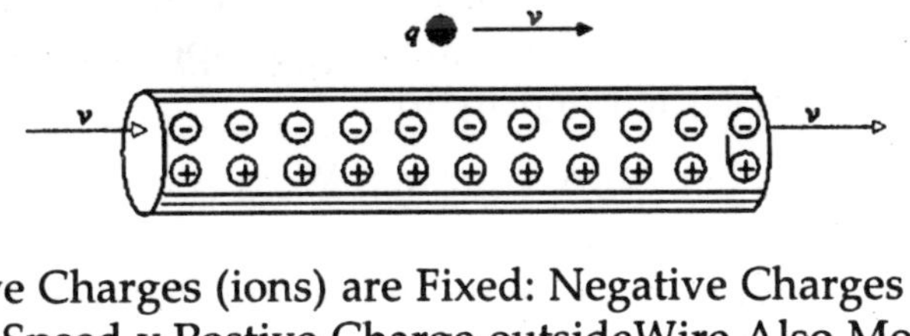

Fig. Postive Charges (ions) are Fixed: Negative Charges (Electrons) Drift at Speed v.Postive Charge outsideWire Also Moves at v.

What force does the positive charge q feel?

The wire is electrically neutral, since it contains equal densities of positive and negative charges, both uniformly distributed throughout the wire. So q feels no electrical force.

However, since q is moving, it will feel a magnetic force,

$$\vec{F}_{mag} = \overrightarrow{qv} \times \vec{B}.$$

The magnetic field lines are vertical circles centreed along the axis of the wire, the field being into the page where the charge is and of magnitude

$$B = \frac{g\mu_0^{1}}{2\pi r} = \frac{\mu_0 \lambda v}{2\pi r}$$

so the force on the charge is of magnitude,

$$F = \frac{qg\mu_0\lambda v^2}{2\pi r}$$

and is directed away from the wire, so the charge will accelerate away from the wire.

Now let us examine the same physical system in the frame of reference in which the charge is initially at rest. In that frame, the electrons are also at rest, but the positive background charge is flowing at v:

Since q is at rest, it cannot feel a magnetic force: such forces depend linearly on speed!

Yet it looks as if it can't feel an electric force either, because the positive and negative charges in the wire have equal densities, right? This leads to the conclusion that q feels no force at all in this frame, so it won't accelerate away from the wire, as it did in the other frame.

The mistake was in ignoring the relativistic Fitzgerald-Lorentz contraction, even though the velocities involved are *millimeters per second*! In the frame in which the wire is at rest, the positive and negative charge densities exactly balance, otherwise there will be extra electrostatic fields that the electrons will quickly move to neutralize.

However, this *necessarily* means that the densities cannot balance exactly in the frame in which the drifting electrons are at rest. In the electrons' rest frame, the positive charges, which had density λ coulombs per meter in their rest frame, are moving at speed v, so relativistic contraction will increase their density to:

$$\frac{\lambda}{\sqrt{1-v^2/c^2}} \cong \lambda + \frac{\lambda v^2}{2c^2}$$

On the other hand, in this electron frame the electrons are at rest, so their density is actually less than it was in the frame of the wire, it has decreased by $\lambda v^2/2c^2$.

The net effect is that the wire, electrically neutral in the lab frame, has a positive charge density $\lambda v^2/c^2$ in the frame of the moving electrons (and the outside charge).

$$\int \vec{E}.d\vec{A} = (\text{enclosed charge}) / \varepsilon_0.$$

Exercise: use this to verify that the electric field from an infinite line of charge λ coulombs per meter has magnitude $\lambda/2\pi r\mu_0$ at distance r from the line charge. From the exercise result, the electrostatic force on the charge q is:

$$F = qE = q\lambda\frac{v^2}{c^2}\frac{1}{2\pi r\varepsilon_0} = \frac{q\lambda v^2}{2\pi r}\frac{1}{\varepsilon_0 c^2} = \frac{q\lambda v^2}{2\pi r}\mu_0$$

where in the last step we used $1/c^2 = \mu_0\varepsilon_0$.

This purely electrical force is identical in magnitude to the purely magnetic force in the other frame! So observers in the two frames *will* agree on the rate at which the particle accelerates away from the wire, but one will call the accelerating force magnetic, the other electric. We are forced to the conclusion that whether a particular force on an actual particle is magnetic or electric, or some mixture of both, depends on the frame of reference—so the distinction is rather artificial.

A much more complete analysis of the fields from moving charges can be found in Purcell, Electricity and Magnetism, McGraw Hill, 1985. We have chosen the simplest possible example to illustrate the basic point that electric and magnetic fields are frame-dependent concepts.

Remarks on General Relativity

Einstein's Parable

In Einstein's little book *Relativity: the Special and the General Theory*, he introduces general relativity with a parable. He imagines going into deep space, far away from gravitational fields, where any body moving at steady speed in a straight line will continue in that state for a very long time.

He imagines building a space station out there - in his words, "a spacious chest resembling a room with an observer inside who is equipped with apparatus." Einstein points out that there will be no gravity, the observer will tend to float around inside the room. But now a rope is attached to a hook in the middle of the lid of this "chest" and an unspecified "being" pulls on the rope with a constant force. The chest and its contents, including the observer, accelerate "upwards" at a constant rate.

How does all this look to the man in the room? He finds himself moving towards what is now the "floor" and needs to use his leg muscles to stand. If he releases anything, it accelerates towards the floor, and in fact all bodies accelerate at the same rate. If he were a normal human being, he would assume the room to be in a gravitational field, and might wonder why the room itself didn't fall. Just then he would discover the hook and rope, and conclude that the room was suspended by the rope.

Einstein asks: should we just smile at this misguided soul? His answer is no - the observer in the chest's point of view is just as valid as an outsider's. In other words, *being inside* the (from an outside perspective) *uniformly accelerating room is physically equivalent to being in a uniform gravitational field.*

This is the basic postulate of general relativity. Special relativity said that all inertial frames were equivalent. General relativity extends this to accelerating frames, and states their equivalence to frames in which there is a gravitational field. This is called the *Equivalence Principle.*

The acceleration could also be used to cancel an existing gravitational field—for example, inside a freely falling elevator passengers are weightless, conditions are equivalent to those in the unaccelerated space station in outer space. It is important to realise that this equivalence between a gravitational field and acceleration is only possible because the gravitational mass is exactly equal to the inertial mass. There is no way to cancel out electric fields, for example, by going to an accelerated frame, since many different charge to mass ratios are possible.

As physics has developed, the concept of fields has been very valuable in understanding how bodies interact with each other. We visualize the electric field lines coming out from a charge, and know that something is there in the space around the charge which exerts a force on another charge coming into the neighbourhood.

We can even compute the energy density stored in the electric field, locally proportional to the square of the electric field intensity. It is tempting to think that the gravitational field is quite similar—after all, it's another inverse square field.

Evidently, though, this is not the case. If by going to an accelerated frame the gravitational field can be made to vanish, at least locally, it cannot be that it stores energy in a simply defined local way like the electric field.

We should emphasize that going to an accelerating frame can only cancel a *constant* gravitational field, of course, so there is no accelerating frame in which the whole gravitational field of, say, a massive body is zero, since the field necessarily points in different directions in different regions of the space surrounding the body.

Some Consequences of the Equivalence Principle

Consider a freely falling elevator near the surface of the earth, and suppose a laser fixed in one wall of the elevator sends a pulse of light horizontally across to the corresponding point on the opposite wall of the elevator. Inside the elevator, where there are no fields present, the environment is that of an inertial frame, and the light will certainly be observed to proceed directly across the elevator.

Imagine now that the elevator has windows, and an outsider at rest relative to the earth observes the light. As the light crosses the elevator, the elevator is of course accelerating downwards at g, so since the flash of light will hit the opposite elevator wall at precisely the height relative to the elevator at which it began, the outside observer will conclude that the flash of light also accelerates downwards at g.

In fact, the light could have been emitted at the instant the elevator was released from rest, so we must conclude that light falls in an initially parabolic path in a constant gravitational field. Of course, the light is traveling very fast, so the curvature of the path is small! Nevertheless, *the Equivalence Principle forces us to the conclusion that the path of a light beam is bent by a gravitational field.*

The curvature of the path of light in a gravitational field was first detected in 1919, by observing stars very near to the sun during a solar eclipse. The deflection for stars observed very close to the sun was 1.7 seconds of arc, which meant measuring image positions on a photograph to an accuracy of hundredths of a millimeter, quite an achievement at the time.

One might conclude from the brief discussion above that a light beam in a gravitational field follows the same path a Newtonian particle would if it moved at the speed of light. This is true in the limit of small deviations from a straight line in a constant field, but is not true even for small deviations for a spatially varying field, such as the field near the sun the starlight travels through in the eclipse experiment mentioned above.

We could try to construct the path by having the light pass through a series of freely falling (fireproof!) elevators, all falling towards the centre of the sun, but then the elevators are accelerating relative to each other (since they are all falling along *radii*), and matching up the path of the light beam through the series is tricky. If it is done correctly (as Einstein did) it turns out that the angle the light beam is bent through is twice that predicted by a naïve Newtonian theory.

What happens if we shine the pulse of light vertically *down* inside a freely falling elevator, from a laser in the centre of the ceiling to a point in the centre of the floor? Let us suppose the flash of light leaves the ceiling at the instant the elevator is released into free fall. If the elevator has height h, it takes time h/c to reach the floor. This means the floor is moving downwards at speed gh/c when the light hits.

Inside the elevator, by the Equivalence Principle, conditions are identical to those in an inertial frame with no fields present. There is nothing to change the frequency of the light. This implies, however, that to an outside observer, stationary in the earth's gravitational field, the frequency of the light *will* change.

This is because he will agree with the elevator observer on what was the initial frequency f of the light as it left the laser in the ceiling (the elevator was at rest relative to the earth at that moment) so if the elevator operator maintains the light had the same frequency f as it hit the elevator floor, which is moving at gh/c relative to the earth at that instant, the earth observer will say the light has frequency $f(1 + v/c) = f(1+gh/c^2)$, using the Doppler formula for very low speeds.

We conclude from this that light shining downwards in a gravitational field is shifted to a higher frequency. Putting the laser in the elevator floor, it is clear that light shining upwards in a gravitational field is red-shifted to lower frequency. Einstein suggested that this prediction could be checked by looking at characteristic spectral lines of atoms near the surfaces of very dense stars, which should be red-shifted compared with the same atoms observed on earth, and this was confirmed.

This has since been observed much more accurately. An amusing consequence, since the atomic oscillations which emit the radiation are after all just accurate clocks, is that *time passes at different rates at different altitudes*. The US atomic standard clock, kept at 5400 feet in Boulder, gains 5 microseconds per year over an identical clock almost at sea level in the Royal Observatory at Greenwich, England.

Both clocks are accurate to one microsecond per year. This means you would age more slowly if you lived on the surface of a planet with a large gravitational field. Of course, it might not be very comfortable.

General Relativity and the Global Positioning System

Despite what you might suspect, the fact that time passes at different rates at different altitudes has significant practical consequences. An important *everyday* application of general relativity is the Global Positioning System. A GPS unit finds out where it is by detecting signals sent from orbiting satellites at precisely timed intervals. If all the satellites emit signals simultaneously, and the GPS unit detects signals from four different satellites, there will be three relative time delays between the signals it detects.

The signals themselves are encoded to give the GPS unit the precise position of the satellite they came from at the time of transmission. With this information, the GPS unit can use the speed of light to translate the detected time delays into distances, and therefore compute its own position on earth by triangulation.

But this has to be done very precisely! Bearing in mind that the speed of light is about one foot per nanosecond, an error of 100 nanoseconds or so could, for example, put an airplane off the runway in a blind landing. This means the clocks in the satellites timing when the signals are sent out must certainly be accurate to 100 nanoseconds a day.

That is one part in 10^{12}.It is easy to check that both the special relativistic time dilation correction from the speed of the satellite, and the general relativistic gravitational potential correction are much greater than that, so the clocks in the satellites must be corrected appropriately.The satellites go around the earth once every twelve hours, which puts them at a distance of about four earth radii.

In fact, Ash by reports that when the first Cesium clock was put in orbit in 1977, those involved were sufficiently skeptical of general relativity that the clock was not corrected for the gravitational redshift effect. But—just in case Einstein turned out to be right—the satellite was equipped with a synthesizer that could be switched on if necessary to add the appropriate relativistic corrections.

After letting the clock run for three weeks with the synthesizer turned off, it was found to differ from an identical clock at ground level by precisely the amount predicted by special plus general relativity, limited only by the accuracy of the clock. This simple experiment verified the predicted gravitational redshift to about one per cent accuracy! The synthesizer was turned on and left on.

Chapter 13

The Speed of Light

As we shall soon see, attempts to measure the speed of light played an important part in the development of the theory of special relativity, and, indeed, the speed of light is central to the theory.

The first recorded discussion of the speed of light (I think) is in Aristotle, where he quotes Empedocles as saying the light from the sun must take some time to reach the earth, but Aristotle himself apparently disagrees, and even Descartes thought that light traveled instantaneously. Galileo, unfairly as usual, in *Two New Sciences* has Simplicio stating the Aristotelian position,

SIMP. Everyday experience shows that the propagation of light is instantaneous; for when we see a piece of artillery fired at great distance, the flash reaches our eyes without lapse of time; but the sound reaches the ear only after a noticeable interval.

Of course, Galileo points out that in fact nothing about the speed of light can be deduced from this observation, except that light moves faster than sound. He then goes on to suggest a possible way to measure the speed of light. The idea is to have two people far away from each other, with covered lanterns.

One uncovers his lantern, then the other immediately uncovers his on seeing the light from the first. This routine is to be practised with the two close together, so they will get used to the reaction times involved, then they are to do it two or three miles apart, or even further using telescopes, to see if the time interval is perceptibly lengthened.

Galileo claims he actually tried the experiment at distances less than a mile, and couldn't detect a time lag. From this one can certainly deduce that light travels at least ten times faster than sound.

Measuring the Speed of Light with Jupiter's Moons

The first real measurement of the speed of light came about half a century later, by a Danish astronomer, Ole Romer, working at the Paris Observatory. He had made a systematic study of Io, one of the moons of

Jupiter, which was eclipsed by Jupiter at regular intervals, as Io went around Jupiter in a circular orbit at a steady rate.Actually, Romer found, for several months the eclipses lagged more and more behind the expected time, until they were running about eight minutes late, then they began to pick up again, and in fact after about six months were running eight minutes early.

The cycle then repeated itself. Romer realised the significance of the time involved-just over one year. This time period had nothing to do with Io, but was the time between successive closest approaches of earth in its orbit to Jupiter. The eclipses were furthest behind the predicted times when the earth was furthest from Jupiter. The natural explanation was that the light from Io (actually reflected sunlight, of course) took time to reach the earth, and took the longest time when the earth was furthest away. From his observations, Romer concluded that light took about twenty-two minutes to cross the earth's orbit. "For it is now certain from the phenomena of Jupiter's satellites, confirmed by the observations of different astronomers, that light is propagated in succession (Note: I think this means at finite speed) and requires about seven or eight minutes to travel from the sun to the earth." This is essentially the correct value.

Of course, to find the speed of light it was also necessary to know the distance from the earth to the sun. During the 1670's, attempts were made to measure the parallax of Mars, that is, how far it shifted against the background of distant stars when viewed simultaneously from two different places on earth at the same time.

This (very slight) shift could be used to find the distance of Mars from earth, and hence the distance to the sun, since all *relative* distances in the solar system had been established by observation and geometrical analysis. According to Crowe, they concluded that the distance to the sun was between 40 and 90 million miles. Measurements presumably converged on the correct value of about 93 million miles soon after that, because it appears Romer (or perhaps Huygens, using Romer's data a short time later) used the correct value for the distance, since the speed of light was calculated to be 125,000 miles per second, about three-quarters of the correct value of 186,300 miles per second.

This error is fully accounted for by taking the time light needs to cross the earth's orbit to be twenty-two minutes (as Romer did) instead of the correct value of sixteen minutes.

Starlight and Rain

The next substantial improvement in measuring the speed of light took place in 1728, in England. An astronomer James Bradley, sailing on the Thames with some friends, noticed that the little pennant on top of the mast changed position each time the boat put about, even though the wind was steady. He thought of the boat as the earth in orbit, the wind as starlight coming from some distant star, and reasoned that the apparent direction the starlight was "blowing" in would depend on the way the earth was moving.

Another possible analogy is to imagine the starlight as a steady downpour of rain on a windless day, and to think of yourself as walking around a circular path at a steady pace. The apparent direction of the incoming rain will not be vertically downwards-more will hit your front than your back. In fact, if the rain is falling at, say, 15 mph, and you are walking at 3 mph, to you as observer the rain will be coming down at a slant so that it has a vertical speed of 15 mph, and a horizontal speed towards you of 3 mph.

Whether it is slanting down from the north or east or whatever at any given time depends on where you are on the circular path at that moment. Bradley reasoned that the apparent direction of incoming starlight must vary in just this way, but the angular change would be a lot less dramatic. The earth's speed in orbit is about 18 miles per second, he knew from Romer's work that light went at about 10,000 times that speed.

That meant that the angular variation in apparent incoming direction of starlight was about the magnitude of the small angle in a right-angled triangle with one side 10,000 times longer than the other, about one two-hundredth of a degree. Notice this would have been just at the limits of Tycho's measurements, but the advent of the telescope, and general improvements in engineering, meant this small angle was quite accurately measurable by Bradley's time, and he found the velocity of light to be 185,000 miles per second, with an accuracy of about one per cent.

Fast Flickering Lanterns

The problem is, all these astronomical techniques do not have the appeal of Galileo's idea of two guys with lanterns. It would be reassuring to measure the speed of a beam of light between two points on the ground, rather than making somewhat indirect deductions based on apparent slight variations in the positions of stars.

We can see, though, that if the two lanterns are ten miles apart, the time lag is of order one-ten thousandth of a second, and it is difficult to see how to arrange that. This technical problem was solved in France about 1850 by two rivals, Fizeau and Foucault, using slightly different techniques.

In Fizeau's apparatus, a beam of light shone between the teeth of a rapidly rotating toothed wheel, so the "lantern" was constantly being covered and uncovered. Instead of a second lantern far away, Fizeau simply had a mirror, reflecting the beam back, where it passed a second time between the teeth of the wheel.

The idea was, the blip of light that went out through one gap between teeth would only make it back through the same gap if the teeth had not had time to move over significantly during the round trip time to the far away mirror. It was not difficult to make a wheel with a hundred teeth, and to rotate it hundreds of times a second, so the time for a tooth to move over could be arranged to be a fraction of one ten thousandth of a second.

The method worked. Foucault's method was based on the same general idea, but instead of a toothed wheel, he shone the beam on to a rotating mirror. At one point in the mirror's rotation, the reflected beam fell on a distant mirror, which reflected it right back to the rotating mirror, which meanwhile had turned through a small angle. After this second reflection from the rotating mirror, the position of the beam was carefully measured.

This made it possible to figure out how far the mirror had turned during the time it took the light to make the round trip to the distant mirror, and since the rate of rotation of the mirror was known, the speed of light could be figured out. These techniques gave the speed of light with an accuracy of about 1,000 miles per second.

Albert Abraham Michelson

Albert Michelson was born in 1852 in Strzelno, Poland. His father Samuel was a Jewish merchant, not a very safe thing to be at the time.

Purges of Jews were frequent in the neighbouring towns and villages. They decided to leave town. Albert's fourth birthday was celebrated in Murphy's Camp, Calaveras County, about fifty miles south east of Sacramento, a place where five million dollars worth of gold dust was taken from one four acre lot.

Samuel prospered selling supplies to the miners. When the gold ran out, the Michelsons moved to Virginia City, Nevada, on the Comstock lode, a silver mining town. Albert went to high school in San Francisco.

In 1869, his father spotted an announcement in the local paper that Congressman Fitch would be appointing a candidate to the Naval Academy in Annapolis, and inviting applications. Albert applied but did not get the appointment, which went instead to the son of a civil war veteran.

However, Albert knew that President Grant would also be appointing ten candidates himself, so he went east on the just opened continental railroad to try his luck. Unknown to Michelson, Congressman Fitch wrote directly to Grant on his behalf, saying this would really help get the Nevada Jews into the Republican party.

This argument proved persuasive. In fact, by the time Michelson met with Grant, all ten scholarships had been awarded, but the President somehow came up with another one. Of the incoming class of ninety-two, four years later twenty-nine graduated. Michelson placed first in optics, but twenty-fifth in seamanship.

The Superintendent of the Academy, Rear Admiral Worden, who had commanded the Monitor in its victory over the Merrimac, told Michelson: "If in the future you'd give less attention to those scientific things and more to your naval gunnery, there might come a time when you would know enough to be of some service to your country."

Sailing the Silent Seas: Galilean Relativity

Michelson was ordered aboard the USS *Monongahela*, a sailing ship, for a voyage through the Carribean and down to Rio. According to the biography of Michelson written by his daughter (*The Master of Light*, by Dorothy Michelson Livingston, Chicago) he thought a lot as the ship glided across the quiet Carribean about whether one could decide in a closed room inside the ship whether or not the vessel was moving. In fact, his daughter quotes a famous passage from Galileo on just this point:

Shut yourself up with some friend in the largest room below decks of some large ship and there procure gnats, flies, and other such small winged creatures. Also get a great tub full of water and within it put certain fishes; let also a certain bottle be hung up, which drop by drop lets forth its water into another narrow-necked bottle placed underneath.

Then, the ship lying still, observe how those small winged animals fly with like velocity towards all parts of the room; how the fish swim indifferently towards all sides; and how the distilling drops all fall into the bottle placed underneath. And casting anything toward your friend, you need not throw it with more force one way than another, provided the distances be equal; and leaping with your legs together, you will reach as far one way as another.

Having observed all these particulars, though no man doubts that, so long as the vessel stands still, they ought to take place in this manner, make the ship move with what velocity you please, so long as the motion is uniform and not fluctuating this way and that.

You will not be able to discern the least alteration in all the forenamed effects, nor can you gather by any of them whether the ship moves or stands still in throwing something to your friend you do not need to throw harder if he is towards the front of the ship from you... the drops from the upper bottle still fall into the lower bottle even though the ship may have moved many feet while the drop is in the air. Of this correspondence of effects the cause is that the ship's motion is common to all the things contained in it and to the air also; I mean if those things be shut up in the room; but in case those things were above the deck in the open air, and not obliged to follow the course of the ship, differences would be observed smoke would stay behind....

Michelson Measures the Speed of Light

On returning to Annapolis from the cruise, Michelson was commissioned Ensign, and in 1875 became an instructor in physics and chemistry at the Naval Academy, under Lieutenant Commander William Sampson. Michelson met Mrs. Sampson's niece, Margaret Heminway, daughter of a very successful Wall Street tycoon, who had built himself a granite castle in New Rochelle, NY. Michelson married Margaret in an Episcopal service in New Rochelle.

At work, demonstrations had just been introduced at Annapolis. Sampson suggested that it would be a good demonstration to measure the speed of light by Foucault's method. Michelson soon realised, on putting together the apparatus, that he could redesign it for much greater accuracy, but that would need money well beyond that available in the teaching demonstration budget.

He went and talked with his father in law, who agreed to put up $2,000. Instead of Foucault's 60 feet to the far mirror, Michelson had about 2,000 feet along the bank of the Severn, a distance he measured to one tenth of an inch. He invested in very high quality lenses and mirrors to focus and reflect the beam.

His final result was 186,355 miles per second, with possible error of 30 miles per second or so. This was twenty times more accurate than Foucault, made the New York Times, and Michelson was famous while still in his twenties. In fact, this was accepted as the most accurate measurement of the speed of light for the next forty years, at which point Michelson measured it again.

Chapter 14

The Michelson-Morley Experiment

THE NATURE OF LIGHT

The speed of light was known to be 186,350 miles per second with a likely error of around 30 miles per second. This measurement, made by timing a flash of light travelling between mirrors in Annapolis, agreed well with less direct measurements based on astronomical observations. Still, this did not really clarify the *nature* of light.

Two hundred years earlier, Newton had suggested that light consists of tiny *particles* generated in a hot object, which spray out at very high speed, bounce off other objects, and are detected by our eyes. Newton's arch-enemy Robert Hooke, on the other hand, thought that light must be a kind of *wave motion*, like sound. To appreciate his point of view, let us briefly review the nature of sound.

The Wavelike Nature of Sound

Actually, *sound* was already quite well understood by the ancient Greeks. The essential point they had realised is that sound is generated by a vibrating material object, such as a bell, a string or a drumhead. Their explanation was that the vibrating drumhead, for example, alternately pushes and pulls on the air directly above it, sending out waves of compression and decompression (known as rarefaction), like the expanding circles of ripples from a disturbance on the surface of a pond.

On reaching the ear, these waves push and pull on the eardrum with the same frequency (that is to say, the same number of pushes per second) as the original source was vibrating at, and nerves transmit from the ear to the brain both the intensity (loudness) and frequency (pitch) of the sound.

There are a couple of special properties of sound waves (actually any waves) worth mentioning at this point. The first is called *interference*. This is most simply demonstrated with water waves. If you put two fingers in a tub of water, just touching the surface a foot or so apart, and vibrate them at the same rate to get two expanding circles of ripples, you will notice that where the ripples overlap there are quite complicated patterns of waves formed.

The essential point is that at those places where the wave-crests from the two sources arrive at the same time, the waves will work together and the water will be very disturbed, but at points where the crest from one source arrives at the same time as the wave trough from the other source, the waves will cancel each other out, and the water will hardly move. You can hear this effect for sound waves by playing a constant note through stereo speakers.

As you move around a room, you will hear quite large variations in the intensity of sound. Of course, reflections from walls complicate the pattern. This large variation in volume is *not* very noticeable when the stereo is playing music, because music is made up of many frequencies, and they change all the time. The different frequencies, or notes, have their quiet spots in the room in different places.

The other point that should be mentioned is that high frequency tweeter-like sound is much more *directional* than low frequency woofer-like sound. It really doesn't matter where in the room you put a low-frequency woofer-the sound seems to be all around you anyway.

On the other hand, it is quite difficult to get a speaker to spread the high notes in all directions. If you listen to a cheap speaker, the high notes are loudest if the speaker is pointing right at you. A lot of effort has gone into designing tweeters, which are small speakers especially designed to broadcast high notes over a wide angle of directions.

Light is a Wave

Bearing in mind the above minireview of the properties of waves, let us now reconsider the question of whether light consists of a stream of particles or is some kind of wave. The strongest argument for a particle picture is that light travels in straight lines. You can hear around a corner, at least to some extent, but you certainly can't see.

Furthermore, no wave-like interference effects are very evident for light. Finally, it was long known, as we have mentioned, that sound waves were compressional waves in air. If light is a wave, just what is waving? It clearly isn't just air, because light reaches us from the sun, and indeed from stars, and we know the air doesn't stretch that far, or the planets would long ago have been slowed down by air resistance.

Despite all these objections, it was established around 1800 that light *is* in fact some kind of wave. The reason this fact had gone undetected for so long was that the wavelength is *really* short, about one fifty-thousandth of an inch.

In contrast, the shortest wavelength sound detectable by humans has a wavelength of about half an inch. The fact that light travels in straight lines is in accord with observations on sound that the higher the frequency (and shorter the wavelength) the greater the tendency to go in straight lines. Similarly, the interference patterns mentioned above for sound waves or ripples on a pond vary over distances of the same sort of size as the wavelengths involved.

Patterns like that would not normally be noticeable for light because they would be on such a tiny scale. In fact, it turns out, there *are* ways to see interference effects with light. A familiar example is the many colours often visible in a soap bubble. These come about because looking at a soap bubble you see light reflected from both sides of a very thin film of water—a thickness that turns out to be comparable to the wavelength of light.

The light reflected from the lower layer has to go a little further to reach your eye, so that light wave must wave an extra time or two before getting to your eye compared with the light reflected from the top layer. What you actually *see* is the *sum* of the light reflected from the top layer and that reflected from the bottom layer.

Thinking of this now as the sum of two sets of waves, the light will be bright if the crests of the two waves arrive together, dim if the *crests* of waves reflected from the top layer arrive simultaneously with the *troughs* of waves reflected from the bottom layer. Which of these two possibilities actually occurs for reflection from a particular bit of the soap film depends on just how much further the light reflected from the lower surface has to travel to reach your eye compared with light from the upper surface, and that depends on the angle of reflection and the thickness of the film.

Suppose now we shine *white* light on the bubble. White light is made up of all the colours of the rainbow, and these different colours have different wavelengths, so we see colours reflected, because for a particular film, at a particular angle, some colours will be reflected brightly.

Light is a Wave, What is Waving?

Having established that light is a wave, though, we still haven't answered one of the major objections raised above. Just what is waving? We discussed sound waves as waves of compression in air.

Actually, that is only one case—sound will also travel through liquids, like water, and solids, like a steel bar. It is found experimentally that, other things being equal, sound travels faster through a medium that is harder to compress—the material just springs back faster and the wave moves through more rapidly. For media of equal springiness, the sound goes faster through the less heavy medium, essentially because the same amount of springiness can push things along faster in a lighter material.

So when a sound wave passes, the material—air, water or solid—waves as it goes through.

Taking this as a hint, it was natural to suppose that light must be just waves in some mysterious material, which was called the *aether*, surrounding and permeating everything. This aether must also fill all of space, out to the stars, because we can see them, so the medium must be there to carry the light. (We could never *hear* an explosion on the moon, however loud, because there is no air to carry the sound to us.) Let us think a bit about what properties this aether must have.

Since light travels so fast, it must be very light, and very hard to compress. Yet, as mentioned above, it must allow solid bodies to pass through it freely, without aether resistance, or the planets would be slowing down. Thus we can picture it as a kind of ghostly wind blowing through the earth. But how can we prove any of this? Can we detect it?

Detecting the Aether Wind

Detecting the aether wind was the next challenge Michelson set himself after his triumph in measuring the speed of light so accurately. Naturally, something that allows solid bodies to pass through it freely is a little hard to get a grip on. But Michelson realised that, just as the speed of sound is relative to the air, so the speed of light must be relative to the aether.

This must mean, if you could measure the speed of light accurately enough, you could measure the speed of light travelling upwind, and compare it with the speed of light travelling downwind, and the difference of the two measurements should be twice the windspeed. Unfortunately, it wasn't that easy.

All the recent accurate measurements had used light travelling to a distant mirror and coming back, so if there was an aether wind along the direction between the mirrors, it would have opposite effects on the two parts of the measurement, leaving a very small overall effect. There was no technically feasible way to do a one-way determination of the speed of light.

At this point, Michelson had a very clever idea for detecting the aether wind. As he explained to his children, it was based on the following puzzle. Suppose we have a river of width w, and two swimmers who both swim at the same speed v feet per second. The river is flowing at a steady rate.

The swimmers race in the following way: they both start at the same point on one bank. One swims directly across the river to the closest point on the opposite bank, then turns around and swims back. The other stays on one side of the river, swimming upstream a distance (measured along the bank) exactly equal to the width of the river, then swims back to the start. Who wins?

Let's consider first the swimmer going upstream and back. Going 100 feet upstream, the speed relative to the bank is only 2 feet per second, so that takes 50 seconds. Coming back, the speed is 8 feet per second, so it takes 12.5 seconds, for a total time of 62.5 seconds.

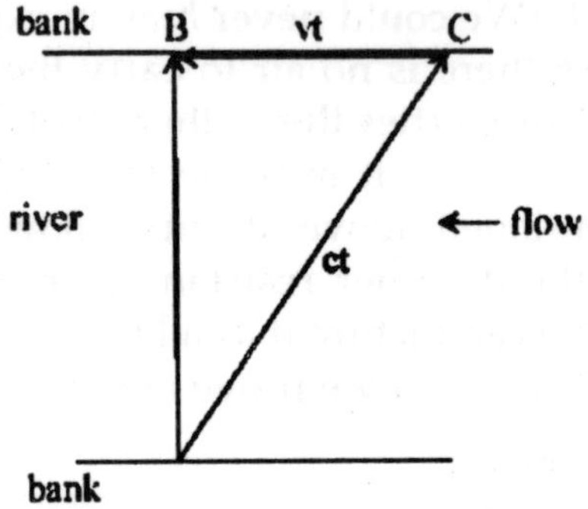

In time t, the swimmer has moved ct relative to the water, and been carried downstream a distance vt. The swimmer going across the flow is trickier. It won't do simply to aim directly for the opposite bank-the flow will carry the swimmer downstream. To succeed in going directly across, the swimmer must actually aim upstream at the correct angle (of course, a real swimmer would do this automatically).

Thus, the swimmer is going at 5 feet per second, at an angle, relative to the river, and being carried downstream at a rate of 3 feet per second. If the angle is correctly chosen so that the net movement is directly across, in one second the swimmer must have moved *four feet* across—the distances covered in one second will form a 3,4,5 triangle. So, at a crossing rate of 4 feet per second, the swimmer gets across in 25 seconds, and back in the same time, for a total time of 50 seconds. The cross-stream swimmer wins. This turns out to true whatever their swimming speed. (Of course, the race is only possible if they can swim faster than the current!)

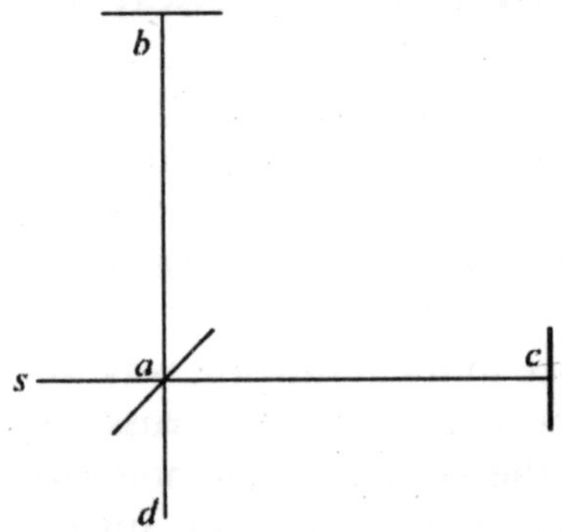

This diagram is from the original paper. The source of light is at *s*, the 45 degree line is the half-silvered mirror, *b* and *c* are mirrors and d the observer. Michelson's great idea was to construct an exactly similar race for pulses of light, with the aether wind playing the part of the river.

The scheme of the experiment is as follows: a pulse of light is directed at an angle of 45 degrees at a half-silvered, half transparent mirror, so that half the pulse goes on through the glass, half is reflected. These two half-pulses are the two swimmers. They both go on to distant mirrors which reflect them back to the half-silvered mirror. At this point, they are again half reflected and half transmitted, but a telescope is placed behind the half-silvered mirror as shown in the figure so that half of each half-pulse will arrive in this telescope. Now, if there is an aether wind blowing, someone looking through the telescope should see the halves of the two half-pulses to arrive at slightly different times, since one would have gone more upstream and back, one more across stream in general. To maximize the effect, the whole apparatus, including the distant mirrors, was placed on a large turntable so it could be swung around. Let us think about what kind of time delay we expect to find between the arrival of the two half-pulses of light.

Taking the speed of light to be c miles per second relative to the aether, and the aether to be flowing at v miles per second through the laboratory, to go a distance w miles upstream will take $w/(c-v)$ seconds, then to come back will take $w/(c+v)$ seconds. The total roundtrip time upstream and downstream is the sum of these, which works out to be $2wc/(c^2-v^2)$, which can also be written $(2w/c).1/(1-v^2/c^2)$.

Now, we can safely assume the speed of the aether is much less than the speed of light, otherwise it would have been noticed long ago, for example in timing of eclipses of Jupiter's satellites. This means v^2/c^2 is a very small number, and we can use some handy mathematical facts to make the algebra a bit easier. First, if x is very small compared to 1, $1/(1-x)$ is very close to $1+x$. (You can check it with your calculator.) Another fact we shall need in a minute is that for small x, the square root of $1+x$ is very close to $1+x/2$. Anyway, the roundtrip upstream-downstream time can be taken, to an excellent approximation, to be $(2w/c).(1+v^2/c^2)$.

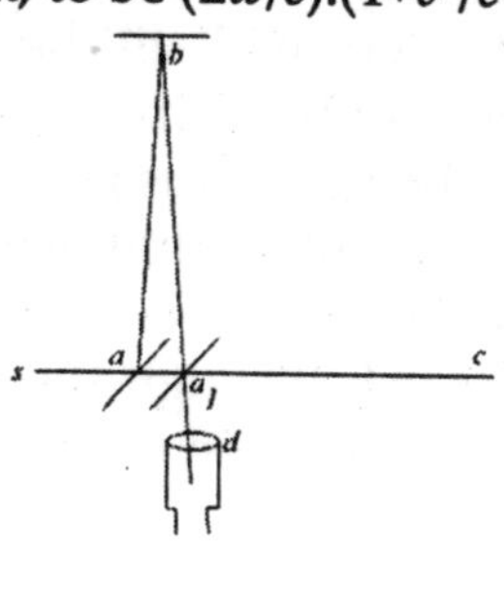

This is also from the original paper, and shows the expected path of light relative to the aether with an aether wind blowing.

Now what about the cross-stream time? The actual cross-stream speed must be figured out as in the example above using a right-angled triangle, with the hypoteneuse equal to the speed c, the shortest side the aether flow speed v, and the other side the cross-stream speed we need to find the time to get across.

From Pythagoras' theorem, then, the cross-stream speed is the square root of (c^2-v^2). Since this will be the same both ways, the roundtrip cross-stream time will be $2w/\text{sqrt}(c^2-v^2)$. This can be written in the form $(2w/c).1/\text{sqrt}(1-v^2/c^2)$, which we shall approximate as $(2w/c).1/(1-v^2/2c^2)$, using the remark about square roots above, and then, replacing $1/(1-x)$ by $1+x$, we finally write the cross-stream roundtrip time as $(2w/c).(1+v^2/2c^2)$.

Looking at the two roundtrip times at the ends of the two paragraphs above, we see that they differ by an amount $(2w/c).v^2/2c^2$. Now, $2w/c$ is just the time the light would take if there were no aether wind at all, say, a few millionths of a second. If we take the aether windspeed to be equal to the earth's speed in orbit, for example, v/c is about 1/10,000, so v^2/c^2 is about 1/100,000,000.

This means the time delay between the pulses reflected from the different mirrors reaching the telescope is about one-hundred-millionth of a few millionths of a second. It seems completely hopeless that such a short time delay could be detected. However, this turns out *not* to be the case, and Michelson was the first to figure out how to do it. The trick is to use the *interference* properties of the lightwaves. Instead of sending pulses of light, Michelson sent in a steady beam of light of a single colour.

This can be visualized as a sequence of ingoing waves, with a wavelength one fifty-thousandth of an inch or so. Now this sequence of waves is split into two, and reflected as previously described. One set of waves goes upstream and downstream, the other goes across stream and back.

Finally, they come together into the telescope and the eye. If the one that took longer is half a wavelength behind, its troughs will be on top of the crests of the first wave, they will cancel, and nothing will be seen. If the delay is less than that, there will still be some dimming. However, slight errors in the placement of the mirrors would have the same effect.

This is one reason why the apparatus is built to be rotated. On turning it through 90 degrees, the upstream-downstream and the cross-stream waves change places. Now the other one should be behind. Thus, if there is an aether wind, if you watch through the telescope while you rotate the turntable, you should expect to see variations in the brightness of the incoming light.

To magnify the time difference between the two paths, in the actual experiment the light was reflected backwards and forwards several times, like a several lap race. Michelson calculated that an aether windspeed of only one or two miles a second would have observable effects in this experiment, so if the aether windspeed was comparable to the earth's speed in orbit around the sun, it would be easy to see. In fact, *nothing* was observed.

The light intensity did not vary at all. Some time later, the experiment was redesigned so that an aether wind caused by the earth's daily rotation could be detected. Again, nothing was seen. Finally, Michelson wondered if the aether was somehow getting stuck to the earth, like the air in a below-decks cabin on a ship, so he redid the experiment on top of a high mountain in California. Again, no aether wind was observed. It was difficult to believe that the aether in the immediate vicinity of the earth was stuck to it and moving with it, because light rays from stars would deflect as they went from the moving faraway aether to the local stuck aether.

The only possible conclusion from this series of very difficult experiments was that the whole concept of an all-pervading aether was wrong from the start. Michelson was very reluctant to think along these lines. In fact, new theoretical insight into the nature of light had arisen in the 1860's from the brilliant theoretical work of Maxwell, who had written down a set of equations describing how electric and magnetic fields can give rise to each other. He had discovered that his equations predicted there could be waves made up of electric and magnetic fields, and the speed of these waves, deduced from experiments on how these fields link together, would be 186,300 miles per second. This is, of course, the speed of light, so it is natural to assume that light is made up of fast-varying electric and magnetic fields. But this leads to a big problem: Maxwell's equations predict a definite speed for light, and it *is* the speed found by measurements.

But what is the speed to be measured relative to? The whole point of bringing in the aether was to give a picture for light resembling the one we understand for sound, compressional waves in a medium. The speed of sound through air is measured relative to air. If the wind is blowing towards you from the source of sound, you will hear the sound sooner. If there isn't an aether, though, this analogy doesn't hold up. So what does light travel at 186,300 miles per second relative to?

There is another obvious possibility, which is called the emitter theory—the light travels at 186,300 miles per second relative to the source of the light. The analogy here is between light emitted by a source and bullets emitted by a machine gun.

The bullets come out at a definite speed (called the muzzle velocity) relative to the barrel of the gun. If the gun is mounted on the front of a tank, which is moving forward, and the gun is pointing forward, then relative to the ground the bullets are moving faster than they would if shot from a tank at rest.

The simplest way to test the emitter theory of light, then, is to measure the speed of light emitted in the forward direction by a flashlight moving in the forward direction, and see if it exceeds the known speed of light by an amount equal to the speed of the flashlight. Actually, this kind of direct test of the emitter theory only became experimentally feasible in the nineteen-sixties.

It is now possible to produce particles, called neutral pions, which decay each one in a little explosion, emitting a flash of light. It is also possible to have these pions moving forward at 185,000 miles per second when they self destruct, and to catch the light emitted in the forward direction, and clock its speed.

It is found that, despite the expected boost from being emitted by a very fast source, the light from the little explosions is going forward at the usual speed of 186,300 miles per second. In the last century, the emitter theory was rejected because it was thought the appearance of certain astronomical phenomena, such as double stars, where two stars rotate around each other, would be affected. Those arguments have since been criticized, but the pion test is unambiguous.

Einstein's View

The results of the various experiments discussed above seem to leave us really stuck. Apparently light is not like sound, with a definite speed relative to some underlying medium. However, it is also not like bullets, with a definite speed relative to the source of the light. Yet when we measure its speed we always get the same result.

Chapter 15

Particles and Waves

FROM BOHR'S ATOM TO ELECTRON WAVES

Reactions to Bohr's Model

Bohr's interpretation of the Balmer formula in terms of quantized angular momentum was certainly impressive, but his atomic model didn't make much mechanical sense, as he himself conceded. For example, an electron jumping from the nth orbit to the mth emitted radiation at frequency equal to the energy difference of the orbits divided by h. Presumably, it began radiating as soon as it left its original orbit.

As Rutherford put it in a letter to Bohr, "how does an electron decide what frequency it is going to vibrate at when it passes from one stationary state to another? It seems to me that you would have to assume that the electron knows beforehand where it is going to stop."

A few days later, Bohr sent Rutherford another manuscript containing the correspondence principle argument gone. This was a bit too long for Rutherford. He responded: "As you know, it is the custom in England to put things very shortly and tersely in contrast to the Germanic method, where it appears to be a virtue to be as long-winded as possible". The Germans themselves were very skeptical of Bohr's model. Harald Bohr wrote to his brother from Göttingen in the fall of 1913 that the young physicists there considered Bohr's model too "bold" and "fantastic".

Mysterious Spectral Lines

One good reason for the Germanic skepticism was the recent discovery of apparently new spectral lines for hydrogen corresponding to *half*-integers in the Balmer formula. These new lines had been seen by a spectroscopist, Alfred Fowler, in a discharge tube containing a mixture of hydrogen and helium, and also in the spectra of a star by Pickering.

$$R_H = \left(\frac{1}{4\pi\varepsilon_0}\right)^2 \cdot \frac{2\pi^2 me^4}{ch^3}$$

When he heard about these new lines, Bohr realized that his formula for the Rydberg constant, should also apply to the ionized helium atom-that is, a helium nucleus with a single electron in orbit around it, provided e^2 is replaced by $2e^2$, to correct for the doubly charged nucleus. This gives an overall factor of 4, exactly equivalent to replacing the integers in the denominator by half-integers. In other words, Bohr argued, the supposed new hydrogen lines in fact were from ionized helium.

Fowler, however, was a very precise experimentalist. He could measure the spectral lines to five significant figures. Therefore, despite uncertainties in the values of e, m and h, the ratio of the "Rydberg constant" for helium to that for hydrogen could be measured to great accuracy. Fowler found it wasn't 4, but actually 4.0016. Bohr's response to this news was conclusive. He pointed out that his analysis of the circular motion had neglected the finite mass of the nucleus. He should really have taken the electron to have an effective mass equal to $mM/(m + M)$. It is easy to verify that if this correction is made for both hydrogen and helium, the ratio changes from 4 to 4.0016!

This discovery won over most physicists to the point of view that Bohr was definitely on to something. Sir James Jeans remarked that the only justification for Bohr's postulates was "the very weighty one of success". According to Rosenfeld (Introduction to *Niels Bohr Collected Works*):

"At Göttingen, that high place of mathematics and physics, where the sense for propriety was strong, the prevailing impression was one of scandal, or at least bewilderment, before the undeserved success of such high-handed disregard of the canons of formal logic." But all Germans didn't feel that way. Bohr's friend George Hevesy told Einstein about Bohr's identification of the Pickering-Fowler lines with helium. Einstein said: "This is an *enormous achievement*. The theory of Bohr must then be right."

A PERIODIC TABLE PUZZLE AND THE X-RAY CONNECTION

Mendeleev's success in demonstrating that the elements had recurring patterns of chemical behaviour when arranged in order of increasing atomic weight had some anomalies. In particular, cobalt has a higher atomic weight than nickel, yet its chemical properties strongly suggested it should come *before* nickel in the table.

It was gradually realized over the period from 1907 to 1913 (primarily by a Dutchman, van den Broek, actually a real estate lawyer who did physics in his spare time) that the significant parameter from a chemical point of view was not the mass number A, but the total number of electrons, and hence the charge on the nucleus Z, once the nuclear model was accepted.

The problem was measuring the nuclear charge. This could in fact be done by Rutherford scattering, but that was a long and tedious process. X-rays gave a quicker way. Any substance on being bombarded with sufficiently fast electrons emits a continuum of x-ray frequencies up to a maximum frequency f given by hf = kinetic energy of electron, *plus some sharply defined lines*. The frequency of the sharply defined lines gradually increases with atomic number.

These x-ray lines were first discovered by Max von Laue in Munich in the summer of 1912. Harry Moseley, who worked with Rutherford mostly teaching undergraduates, decided to investigate this new spectroscopy. When Bohr came to visit in 1913, he encouraged Moseley to investigate x-ray spectra of many elements with a view to pinning down the values of Z. Moseley used a cathode ray tube a yard long with a little train inside it having a different element in each car, so many readings could be taken in one run.

He found several different x-ray lines for each element, with very similar patterns for each element, although gradually shifting to higher frequencies with increasing Z. For example, one particular line, labeled *Ka*, had a Z-dependent frequency:

$$f(Z) = R_H(Z-1)^2\left(\frac{1}{1^2}-\frac{1}{2^2}\right)$$

This made it possible to order all the elements unambiguously. It was also good evidence that Bohr's theory was close to correct for the innermost electrons in the atom, those in the field of the unshielded nucleus. Actually, the appearance of Z – 1 rather than Z in the formula is because the innermost shell has just two electrons, and they partially shield the nuclear charge from each other. The appearance of this factor was not understood at the time, because it was thought that there were more electrons in the innermost shell.

Bohr later remarked that Moseley's extremely clear x-ray work had far more impact than Rutherford's scattering experiment in widening acceptance of the Bohr-Rutherford atom. Tragically, Moseley died two years later, killed by a bullet in the head at Gallipoli.

What Bohr's Model Achieved

As Bohr himself emphasized, his model atom didn't *explain* anything, in the way classical dynamics or electromagnetism could explain how systems functioned in terms of a few underlying laws. What the Bohr model did do was to tie together previously unrelated phenomena, and therefore clarify the agenda for future investigations. It showed there was a link between the Balmer formula, Planck's constant and the nuclear atom, with quantized angular momentum playing a central role.

The precision of the spectral predictions for *ionized* helium demonstrated something was right about the model, as did the interpretation of characteristic x-rays for many elements in terms of inner-shell electrons.

On the other hand, despite heroic efforts over several years by Bohr and others, there was no successful prediction of the observed spectral lines for the next simplest atom neutral helium. Evidently, to make further progress in understanding atoms, some new insight was needed.

At this point the war intervened. Much of the momentum in physics was lost. Rutherford's lab was practically emptied, Rutherford himself spent most of his time on antisubmarine research. Bohr put much effort into establishing a physics institute in Copenhagen. He also thought at length about how the periodic table might be understood in terms of filling an atom with electrons.

There was something of a breakthrough in 1924 when the German physicist Wolfgang Pauli suggested there must be a law that two electrons couldn't go into the same orbit. (By this point, the number of allowed orbits had been substantially increased by Sommerfeld to include Keplerian elliptic ones.) Pauli's "exclusion principle" was introduced as an empirical rule.

We shall see later that it follows very naturally from symmetry considerations. We shall also find that in the full quantum picture of the atom, although the angular momentum is quantized, as Bohr realized, the orbital angular momentum is actually *zero* in the ground state of the hydrogen atom.

One of the more surprising results of the full quantum theory is that the energy levels found by Bohr, and even relativistic corrections computed by Sommerfeld, are exactly correct, even though the model of classical inverse square orbits is far from the truth (except for very large orbits).

A Student Prince Catches a Wave

Despite all the effort in Copenhagen and Göttingen, the next real advance in understanding the atom came from an unlikely quarter a student prince in Paris. Prince Louis de Broglie was a member of an illustrious family, prominent in politics and the military since the 1600's.

Louis began his university studies with history, but his elder brother Maurice studied x-rays in his own laboratory, and Louis became interested in physics. He worked with the very new radio telegraphy during the war.

After the war, de Broglie focused his attention on Einstein's two major achievements, the theory of special relativity and the quantization of light waves. He wondered if there could be some connection between them. Perhaps the quantum of radiation really should be thought of as a particle.

If so, the theory of special relativity suggested that the observed energy-momentum relationship, $E = cp$, meant that it had a very small rest mass, for it was always observednaturallyto be moving at the speed of light.

De Broglie suspected that if the speed of a sufficiently low energy quantum could be measured, it would be found to be less than c. On this point he was wrong (as far as we know!). Nevertheless, it was a very valuable conceptual breakthrough to think of the quantum of radiation as a *particle*, knowing full well that radiation is a wave.

In fact, his incorrect idea that the photon (as we now call the light quantum) had a rest mass led him to analyse the relationship between particle properties and wave properties by transforming to the rest frame of the photon, and he discovered that the energy and momentum of the particle were related to the frequency and wavelength of the wave by:

$$E = hf,\ p = h/\lambda$$

Of course, the first condition is the Planck-Einstein quantization, and the second follows trivially from it if we take $E = cp$ and $\lambda f = c$. But de Broglie showed it was more generally true it worked even if the photon had a rest mass.

Having decided that the photon might well be a particle with a rest mass, even if very small, it dawned on de Broglie that in other respects it might not be too different from other particles, especially the very light electron. In particular, maybe the electron also had an associated wave. The obvious objection was that if the electron was wavelike, why had no diffraction or interference effects been observed?

But there was an answer. If de Broglie's relation between momentum and wavelength, $p = h/\lambda$, also held for electrons, the wavelength was sufficiently short that these effects would be easy to miss. As de Broglie himself pointed out, the wave nature of light isn't very evident in everyday life, or in ray tracing in geometrical optics.

He suspected the apparently pure particle nature of electronic trajectories was analogous to the apparent straight line propagation of rays of light, over distance scales much greater than the wavelength.

However, the wavelike properties should be important on an atomic scale. No progress had been made in a decade in understanding why the electronic orbits in the Bohr atom were restricted to integral values of the angular momentum in units of h. But if the electron were in some sense a wave, it would be very natural to restrict the orbits to those of standing waves, for otherwise the electron wave on going around the orbit would interfere with itself destructively.

Suppose now the electron, having momentum p, is moving in a circular orbit of radius r. Then for a standing wave, a whole number of wavelengths must fit around the circle, so for some integer n, $n\lambda = 2\pi r$.

Putting this together with $p = h/\lambda$, we find:

$$2\pi r = n\lambda = nh/p$$

so,

$$L = pr = nh/2\pi.$$

The "standing wave" condition immediately gives Bohr's quantization of angular momentum!

This was the prince's Ph. D. thesis, presented in 1924. His thesis advisor was somewhat taken aback, and wasn't sure if this was sound work. He asked de Broglie for an extra copy of the thesis, which he sent to Einstein. Einstein wrote shortly afterwards: *"I believe it is a first feeble ray of light on this worst of our physics enigmas"*.

An Accident at the Phone Company Makes Everything Crystal Clear

There was an accident at the Bell Telephone Laboratories in April 1925. Clinton Davisson and L. H. Germer, looking for ways to improve vacuum tubes, were watching how electrons from an electron gun in a vacuum tube scattered off a flat nickel surface. Suddenly, while the experiment was running and the nickel target was very hot, a bottle of liquid air near the apparatus exploded, smashing one of the vacuum pipes, and air rushed into the apparatus.

The hot nickel target oxidized immediately. The layer of oxide made their target useless for further investigations. They decided to clean off the oxide by heating the nickel in a hydrogen atmosphere then in vacuum. After doing this for a prolonged period, the nickel looked good, and they resumed the investigation.

To their amazement, the pattern of electron scattering from the newly cleaned nickel target was completely different from that before the accident. What had changed? On examining their newly cleaned crystal carefully, they found a clue. The original target was polycrystalline made up of a multitude of tiny crystals, oriented randomly. During the prolonged heating of the cleaning process, the nickel had re-crystallized into a few large crystals.

To quote from their paper: "It seemed probable to us from these results that the intensity of scattering from a single crystal would exhibit a marked dependence on crystal direction, and we set about at once preparing experiments for an investigation of this dependence. We must admit that the results obtained in these experiments have proved to be quite at variance with our expectations.

It seemed likely that strong beams would be found issuing from the crystal along what may be termed its transparent directions the directions in which the atoms in the lattice are arranged along the smallest number of lines per unit area. Strong beams are indeed found issuing from the crystal, but only when the speed of bombardment lies near one or another of a series of critical values, ant then in directions quite unrelated to crystal transparency.

"The most striking characteristic of these beams is a one to one correspondence which the strongest of them bear to the Laue beams that would be found issuing from the same crystal if the incident beam were a beam of x-rays. Certain others appear to be analogues of optical diffraction beams from plane reflection gratings the lines of these gratings being lines or rows of atoms in the surface of the crystal.

Because of these similarities a description in terms of an equivalent wave radiation is not only possible, but most simple and natural. This involves the association of a wavelength with the incident electron beam, and this wavelength turns out to be in acceptable agreement with the value h/mv of the undulatory mechanics, Planck's action constant divided by the momentum of the electron.

"That evidence for the wave nature of particle mechanics would be found in the reaction between a beam of electrons and a single crystal was predicted by Elsasser two years agoshortly after the appearance of L. de Broglie's original papers on wave mechanics."

WAVE PACKETS

THE WAVE-PARTICLE PUZZLE

De Broglie's explanation of the Bohr atom quantization rules, together with the accidental discovery of electron diffraction scattering by Davisson and Germer, make a very convincing case for the wave nature of the electron. Yet the electron certainly behaves like a particle sometimes.

An electron has a definite mass and charge, it can move slowly, it can travel through a piece of apparatus from a gun to a screen. What, then, is the relationship between the wave and particle viewpoints? De Broglie himself always felt both were always present. He called the wave a pilot wave, and thought it guided the motion of the particle. Unfortunately, that viewpoint leads to contradictions.

The standard modern interpretation is that the intensity of the wave (measured by the square of its amplitude) at any point gives the relative probability of finding the particle at that point. This interpretation, originally presented by Max Born in 1926, is parallel to the relation between the electromagnetic field and quanta the probability of finding a quantum (photon) at any point is proportional to the energy density of the field at that point, which is the square of the electric field vector plus the square of the magnetic field vector. The standard notation for the de Broglie wave function associated with the electron is y (x,t).

Thus, | y (*x,t*) |^2*Dx* is the relative probability of finding the electron in a small interval of length *Dx* near point *x* at time *t*. (For the moment, we restrict the electron to move in one dimension for simplicity. The generalization is straightforward.)

Keeping the Wave and the Particle Together?

Suppose following de Broglie we write down the relation between the "particle properties" of the electron and its "wave properties":

$$½mv^2 = E = hf,\ mv = p = h/\lambda$$

It would seem that we can immediately figure out the speed of the wave, just using l *f* = *c¢*, say. We find:

$$l\, f = (h/mv).\ (½mv^2/h) = ½v.$$

So the speed of the wave seems to be only half the speed of the electron!

How could they stay together? What's wrong with this calculation?

Localizing an Electron

To answer this question, it is necessary to think a little more carefully about the wave function corresponding to an electron traveling through a vacuum tube, say. The electron leaves the cathode, shoots through the vacuum, and impinges on an anode of a grid.

At an intermediate point in this process, it is moving through the vacuum and the wave function must be nonzero over some volume, but zero in the places the electron has not possibly reached yet, and zero in the places it has definitely left.

However, if the electron has a precise energy, say fifty electron volts, it also has a precise momentum. This necessarily implies that the wave has a precise wavelength. But the only wave with a *precise* wavelength l has the form

$$\psi(x, t) = A \sin (kx - \omega t)$$

where $k = 2\pi/\lambda$, and $\omega = 2\pi f$. The problem is that this plane sine wave extends to infinity in both spatial directions, so cannot represent a particle whose wave function is non zero in a limited region of space.

Therefore, to represent a localized particle, we must superpose waves having different wavelengths. The principle is best illustrated by superposing two waves with slightly different wavelengths, and using the trigonometric addition formula:

$$\sin((k - \Delta k)x - (\omega - \Delta\omega)t) + \sin((k + \Delta k)x - (\omega + \Delta\omega)t)$$

$$= 2\sin(k - \omega k)\cos((\Delta k)x - (\Delta\omega)t)$$

This formula represents the phenomenon of beats between waves close in frequency. The first term, sin(*kx*-w *t*), oscillates at the average of the two frequencies. It is modulated by the slowly varying second term, which oscillates once over a spatial extent of order $\pi/\Delta k$. This is the distance over which waves initially in phase at the origin become completely out of phase. Of course, going a further distance of order $\pi/\Delta k$, the waves will become synchronized again.

That is, beating two close frequencies together breaks up the continuous wave into a series of packets, the beats. To describe a single electron moving through space, we need a single packet.

This can be achieved by superposing waves having a continuous distribution of wavelengths, or wave numbers within of order Δk, say of k. In this case, the waves will be out of phase after a distance of order $\pi/\Delta k$, but since they have many different wavelengths, they will never get back in phase again. (This is all covered properly in the theory of Fourier transforms. Here we are just trying to make it plausible.)

The Uncertainty Principle

It should be evident from the above argument that to construct a wave packet representing an electron localized in a small region of space, the component waves must get out of phase rapidly.

This means their wavelengths cannot be very close together. In fact, it is not difficult to give a semi-quantitative estimate of the spread in wavelength necessary, just from a consideration of the two beating waves.

A packet localized in a region of extent Δx can be constructed of waves having k's spread over a range Δk, where $\Delta x \sim \pi/\Delta k$.

Now, $k = 2\pi/l$, and $p = h/\lambda$, so $k = 2\pi\, p/h$.

Therefore, $\Delta k = 2\pi\, \Delta p/h$, and $\Delta x \sim \pi/\Delta k \sim h/\Delta p$ (dropping the factor of 2).

Thus:

$$\Delta x \Delta p \sim h$$

This is Heisenberg's Uncertainty Principle.

Phase Velocity and Group Velocity: Keeping the Wave and Particle Together

Establishing that an electron moving through space must be represented by a wave packet also resolves the paradox that the velocity of the waves seems to be different from the velocity of the electron. The point is that the electron waves, like water waves but unlike electromagnetic waves, have differing phase and group velocities.

To see this, consider again the beating of two waves of slightly different wavelengths.

$$\sin((k - \Delta k)x - (\omega - \Delta\omega)t) + \sin((k + \Delta k)x - (\omega + \Delta\omega)t) =$$

$$2\sin(kx - \omega t)\cos((\Delta k)x - (\Delta\omega)t)$$

The waves described by the term $\sin(kx - \omega t)$ have velocity $\frac{1}{2}v$, as previously derived. But the envelope, the shape of the wave packet, has velocity $\Delta\omega/\Delta k$ rather than ω/k. These velocities would be the same if ω were linear in k, as it is for ordinary electromagnetic waves. But the $\omega - k$ relationship follows from the energy-momentum relationship for the (non-relativistic) electron, $E = \frac{1}{2}mv^2 = p^2/2m$.

So $dE/dp = p/m = v$. But $E = hf = h\omega/2p$, and $p = h/\lambda = hk/2\pi$.

Therefore, $\Delta\omega/\Delta k = dE/dp = v$. So the packet travels at the speed we know the electron must travel at, even though the wave peaks within the packet travel at one-half the speed.

PROBABILITIES, AMPLITUDES AND PROBABILITY AMPLITUDES

Here we review three kinds of double slit experiments *classical particles, classical waves,* and *quantum things.*

BULLETS

First we consider a double slit experiment with *bullets.* The source is a not very accurate machine gun, which sprays bullets towards two parallel slits, 1 and 2, in a sheet of armorplate, which may be covered one at a time with another piece of armor. The bullets are "detected" by plowing into a soft (but thick) wood screen some distance beyond the slits. We assume the bullets don't break up, you always find whole bullets in the wood.

If we now cover slit 2, and fire a lot of bullets, by examining the screen at the end we will see where the bullets went, and be able to construct a probability distribution $P_1(x)dx$ giving the probability that a bullet going through slit 1 lands in the interval $x, x + dx$. Similarly, covering slit 1, we can find the probability distribution $P_2(x)$ for bullets going through slit 2. Now suppose we open both slits this gives a probability distribution $P_{12}(x)$ and it is clear that this will be simply the sum:

$$P_{12}(x) = P_1(x) + P_2(x).$$

This of course follows from the self-evident fact that any bullet getting to the screen must have gone either through slit 1 or slit 2. *The probabilities add.*

There is one other important point about bullets they are lumpy. You either find one or you don't. This is also true of photons and electrons.

WATER WAVES

Now think about a double slit experiment with waves, say, surface waves on water the standard ripple tank experiment. Assume that plane waves of definite wavelength impinge from the left on a barrier having two slits P_1 and P_2, with P_2 being initially closed. Then to the right of the barrier, waves emanate from the open slit with circular wavefronts. These waves are detected at a screen over to the right. The waves could be detected by having corks bobbing up and down in the water just in front of the screen.

We could at some instant in time measure the height distribution $h_1(x)$ of the corks from waves coming through slit 1 (slit 2 being closed). Of course, this height distribution is changing all the time, and is not really what interests us. What we really want to know is the intensity of the wave motion at each point on the screen (analogous to the brightness of light, for example). In any simple harmonic motion, the intensity is proportional to the square of the amplitude. For our surface waves on water, the amplitude is of course just the height of the wave (relative to still water), so the energy intensity $I_1(x)$ for waves coming through slit 1 has the form

$$I_1(x) = Ah_1^2(x).$$

What happens when both slits are open? It is well known that waves superpose linearly—that is, the *heights* add. If at some instant of time with both slits open the height distribution of the corks is given by $h_{12}(x)$, then

$$h_{12}(x) = h_1(x) + h_2(x).$$

In this harmonic motion, the corks bob downwards as much as upwards, so h_1 and h_2 are as likely to cancel as they are to reinforce each other. The energy intensity when both slits are open,

$$I_{12}(x = Ah_{12}^2(x) = I_1(x) + I_2(x) + Ah_1(x)h_2(x).$$

The crucial point here is that in contrast to the bullets, the intensity when both slits are open is *not* the sum of the separate intensities for the slits being open one at a time. This is just the familiar wave interference effect.

Notice that in this classical wave picture, there is no lumpiness. The energy of the wave lapping against the screen varies smoothly, and can have any value we choose.

CLASSICAL LIGHT

Historically, an experiment equivalent to the double slit experiment for light was what convinced people it was indeed a wave. The light reaching a screen exhibited the identical intensity pattern observed for water waves.

This was fully explained with the advent of Maxwell's equations: the light was simply an electromagnetic wave, the amplitude of the electric field being analogous to the height of the water wave in the discussion above. To find the total electric field E_{12} at any point on the screen, the electric field E_1 of the wave emanating from S_1 is added to E_2 from S_2.

The intensity of light I_{12} reaching the screen is proportional to E_{12}^2. Just as for the water waves, E_1 and E_2 oscillate about mean values of zero, so the analogy is complete.

PHOTONS

The discovery that light is after all lumpy on a fine enough scale throws this picture into confusion. Experimentally, light of frequency f can only be detected in lumps of size hf.

To the classically trained mind, this naturally suggests that the source of light is actually sending out these lumps, and each lump presumably passes through one of the slits on its way to the screen. However, this naïve picture leads to a contradiction—how could it generate the observed diffraction pattern?

Experimentally, the diffraction pattern builds up even if the light is so dim that there is a time interval between each photon (lump) passing through the apparatus. The picture of each photon going through one of the slits cannot be correct, because the distance between the slits governs the brightness pattern on the screen, and thus the probability pattern of where the photons land.

The only known resolution to this conceptual problem is to give up trying to visualize the path of the photon, just solve Maxwell's equation for the wave propagation, then interpret the resulting wave intensity as giving the relative probability of detecting the photon at any point. This is how quantum mechanics is done. It does not correspond to a physical picture readily interpretable in terms of familiar concepts. However, it does accord well with what is observed in nature.

ELECTRONS

Electrons behave in the same way as photons; there is little to add to the description. The irony is, of course, that before quantum mechanics electrons were seen as bullets, light as waves, and it has turned out that they are *both* basic quantum particles, with their propagation governed by *wave* equations: but when they arrive, they act like *particles*.

THE UNCERTAINTY PRINCIPLE

WAVES ARE FUZZY

The *wave nature* of particles implies that we cannot know *both* position *and* momentum of a particle to an arbitrary degree of accuracy if Δx represents the uncertainty in our knowledge of position, and that of momentum, then

$$\Delta p \Delta x \sim h$$

where h is Planck's constant. In the real world, particles are three-dimensional and we should say

$$\Delta p_x \Delta x \sim h$$

with corresponding equations for the other two spatial directions. The fuzziness about position is related to that of momentum *in the same direction*.

Let's see how this works by trying to measure y-position and y-momentum very accurately. Suppose we have a source of electrons, say, an electron gun in a CRT (cathode ray tube, such as an old-fashioned monitor). The beam spreads out a bit, but if we interpose a sheet of metal with a slit of width w, then for particles that make it through the slit, we know y with an uncertainty $\Delta y = w$.

Now, if the slit is a long way downstream from the electron gun source, we also know p_y very accurately as the electron reaches the slit, because to make it to the slit the electron's velocity would have to be aimed just right.

But does the measurement of the electron's y position in other words, having it go through the slit affect its y momentum? The answer is *yes*. If it didn't, then sending a stream of particles through the slit they would all hit very close to the same point on a screen placed further downstream.

But we know from experiment that this is *not* what happens a single slit diffraction pattern builds up, of angular width $\theta \sim \lambda/w$, where the electron's de Broglie wavelength λ is given by $px \cong h/\lambda$ (there is a negligible contribution to λ from the y-momentum). The consequent uncertainty in p_y is

$$\Delta p_y/p_x \sim \theta \sim \lambda/w$$

Putting in $p_x = h/\lambda$, we find immediately that

$$\Delta p_y \sim h/w$$

so the act of measuring the electron's y position has fuzzed out its y momentum by precisely the amount required by the uncertainty principle.

TRYING TO BEAT THE UNCERTAINTY PRINCIPLE

In order to understand the Uncertainty Principle better, let's try to see what goes wrong when we actually try to measure position and momentum more accurately than allowed.

For example, suppose we look at an electron through a microscope. What could we expect to see? Of course, you know that if we try to look at something *really* small through a microscope it gets blurry a small sharp object gets diffraction patterns around its edges, indicating that we are looking at something of size comparable to the wavelength of the light being used.

If we look at something much smaller than the wavelength of light like the electron we would expect a diffraction pattern of concentric rings with a circular blob in the middle The size of the pattern is of order the wavelength of the light, in fact from optics it can be shown to be ~ $\lambda f/d$ where d is the diameter of the object lens of the microscope, f the focal distance (the distance from the lens to the object). We shall take $f/d \sim 1$, as it usually is. So looking at an object the size of an electron should give a diffraction pattern centered on the location of the object. That would seem to pin down its position fairly precisely.

What about the *momentum* of the electron? Here a problem arises that doesn't matter for larger objects the light we see has, of course, bounced off the electron, and so the electron has some recoil momentum. That is, by bouncing light off the electron we have given it some momentum.

Can we say how much? To make it simple, suppose we have good eyes and only need to bounce one photon off the electron to see it. We know the initial momentum of the photon (because we know the direction of the light beam we're using to illuminate the electron) and we know that after bouncing off, the photon hits the object lens and goes through the microscope, but we don't know *where* the photon hit the object lens.

The whole point of a microscope is that all the light from a point, light that hits the object lens *in different places,* is all focused back to one spot, forming the image (apart from the blurriness mentioned above).

So if the light has wavelength λ, its constituent photons have momentum $\sim h/\lambda$, and from our ignorance of where the photon entered the microscope we are uncertain of its x-direction momentum by an amount $\sim h/\lambda$. Necessarily, then, we have the same uncertainty about the electron's x-direction momentum, since this was imparted by the photon bouncing off.

But now we have a problem. In our attempts to minimize the uncertainty in the electron's momentum, by only using one photon to detect it, we are not going to see much of the diffraction pattern discussed above such diffraction patterns are generated by *many* photons hitting the film, retina or whatever detecting equipment is being used. A single photon generates a single point (at best!). This point will most likely be within of order λ of the centre of the pattern, but this leaves us with an uncertainty in position of order λ.

Therefore, in attempting to observe the position and momentum of a single electron using a single photon, we find an uncertainty in position $\Delta x \sim \lambda$, and in momentum $\Delta px \sim h/\lambda$. These results are in accordance with Heisenberg's Uncertainty Principle $\Delta x.\Delta p_x \sim h$.

Of course, we could pin down the position much better if we used N photons instead of a single one. From statistical theory, it is known that the remaining uncertainty $\sim \lambda / \sqrt{N}$. But then N photons have bounced off the electron, so, since each is equally likely to have gone through any part of the object lens, the uncertainly in momentum of the electron as a consequence of these collisions goes up as $\sqrt{N}$. (The same as the average imbalance between heads and tails in a sequence of N coin flips.)

Noting that the uncertainty in the momentum of the electron arises because we don't know where the bounced-off photon passes through the object lens, it is tempting to think we could just use a smaller object lens, that would reduce Δp_x.

Although this is correct, recall from above that we stated the size of the diffraction pattern was $\sim\lambda f/d$, where d is the diameter of the object lens and f its focal length. It is easy to see that the diffraction pattern, and consequently Δx, gets bigger by just the amount that Δp_x gets smaller!

WATCHING ELECTRONS IN THE DOUBLE SLIT EXPERIMENT

Suppose now that in the double slit experiment, we set out to detect which slit each electron goes through by shining a light just behind the screen and watching for reflected light from the electron immediately after it had passed through a slit. Following the discussion in Feynman's *Lectures in Physics*, Volume III, we shall now establish that if we can detect the electrons, we ruin the diffraction pattern!

$$\left(n+\frac{1}{2}\right)\lambda_{\text{elec}} = d \sin \theta.$$

Taking the distance between the two slits to be d, the dark lines in the diffraction pattern are at angles

If the light used to see which slit the electron goes through generates an uncertainty in the electron's y momentum Δp_y, in order not to destroy the diffraction pattern we must have

$$\Delta p_y / p < \lambda_{\text{elec}} / d$$

(the angular uncertainty in the electron's direction must not be enough to spread it from the diffraction pattern maxima into the minima). Here p is the electron's full momentum, $p = h/\lambda_{\text{elec}}$. Now, the uncertainty in the electron's y momentum, looking for it with a microscope, is $\Delta p_y \sim h/\lambda_{\text{light}}$.

Substituting these values in the inequality above we find the condition for the diffraction pattern to survive is

$$\lambda_{\text{light}} > d,$$

the wavelength of the light used to detect which slit the electron went through must be greater than the distance between the slits. Unfortunately, the light scattered from the electron then gives one point in a diffraction pattern of size the wavelength of the light used, so even if we see the flash this does not pin down the electron sufficiently to say which slit it went through. Heisenberg wins again.

UNCERTAINTY PRINCIPLE DETERMINES THE SIZE OF EVERYTHING

It is interesting to see how the actual physical size of the hydrogen atom is determined by the wave nature of the electron, in effect, by the Uncertainty Principle. In the ground state of the hydrogen atom, the electron minimizes its total energy. For a classical atom, the energy would be minus infinity, assuming the nucleus is a point (and very large in any case) because the electron would sit right on top of the nucleus. However, this cannot happen in quantum mechanics.

Such a very localized electron would have a very large uncertainty in momentum in other words, the kinetic energy would be large. This is most clearly seen by imagining that the electron is going in a circular orbit of radius r with angular momentum $h/2\pi$. Then one wavelength of the electron's de Broglie wave just fits around the circle, $\lambda_{elec} = 2\pi r$. Clearly, as we shrink the circle's radius r, λ_{elec} goes down proportionately, and the electrons momentum

$$p = h/\lambda_{elec} = h/2\pi r$$

increases. Adding the electron's electrostatic potential energy we find the total energy for a circular orbit of radius r is:

$$E(r) = K.E. + P.E. = \frac{p^2}{2m} - \frac{e^2}{4\pi\varepsilon_0 r} = \frac{h^2}{8m\pi^2 r^2} - \frac{e^2}{4\pi\varepsilon_0 r}.$$

Notice that for very large r, the potential energy dominates, the kinetic energy is negligible, and shrinking the atom lowers the total energy. However, for small enough r, the (always positive) kinetic energy term wins, and the total energy *grows* as the atom shrinks. Evidently, then, there must be a value of r for which *the total energy is a minimum.* Visualizing a graph of the total energy given by the equation above as a function of r, at the minimum point the slope of $E(r)$ is zero, $dE(r)/dr = 0$.

That is,

$$-\frac{h^2}{4m\pi^2 r} + \frac{e^2}{4\pi\varepsilon_0} = 0$$

giving,

$$r_{min} = \frac{\varepsilon_0 h^2}{\pi m e^2}.$$

The total energy for this radius is the exact right answer, which is reassuring. The point of this exercise is to see that in quantum mechanics, unlike classical mechanics, a particle cannot position itself at the exact minimum of potential energy, because that would require a very narrow wave packet and thus be expensive in kinetic energy.

The ground state of a quantum particle in an attractive potential is a trade off between potential energy minimization and kinetic energy minimization. Thus the physical sizes of atoms, molecules and ultimately ourselves are determined by Planck's constant.

Bibliography

Abrikosov, A. A: *Fundamentals of the Theory of Metals,* 1994.

Ahlfors, L. V: *Complex Analysis,* 1996.

Arfken, G. B: *Mathematical Methods for Physicists,* 1998.

Akhiezer, N. I: *The Calculus of Variations,* 1997.

Andersen, H: *Nordic Solid State Physics Conference,* 1996.

Arnold, V. I: *Ordinary Differential Equations,* 1995.

Ashcroft, N. W: *Solid State Physics,* 1996.

Barrett, L. C: *Advanced Engineering Mathematics,* 1998.

Bartle, R.G: *The elements of Real Analysis,* 1997.

Bazaraa, M. S: *Nonlinear Programming (Theory and Algorithms),* 1999.

Bedient, P. E: *Elementary Differential Equations,* 2000.

Berry, M: *Principles of Cosmology and Gravitation,* 1998.

Birkhoff, G: *Ordinary Differential Equations,* 2002.

Blom, Gunnar: *Probability and Statistics (Theory and Applications),* 1999.

Boothby, W. M: *An Introduction to Differentiable Maniflods and Riemannian Geometry,* 1996.

Bransden, B. H: *Physics of atoms and molecules,* 1997.

Brink, D. M: *Angular Momentum,* 2001.

Brown, E. B: *Modern Optics,* 1995.

Buck, R. C: *Advanced Calculus,* 1998.

Bush, B.W: *Empirical Characterization of Infrastructure Networks,* 1995.

Chevally, C: *Theory of Lie groups I,* 1996.

Chorin, A. J: *A Mathematical Introduction to Fluid Mechanics*, 1999.

Coddington, E. A: *The Theory of Ordinary Differential Equations*, 1995.

Cole, RJ: *Essential Solid State Physics: Essential Solid State Physics*, 2000.

Dalven, R: *Introduction to applied solid state physics*, 1995.

Dupasquier, A: *Positron solid-state physics: Proceedings of the International*, 1995.

Fano, R. M: *Electromagnetic Fields, Energy, and Forces*, 1999.

Edmonds, A. R: *Angular Momentum in Quantum Mechanics*, 1997.

Emily, D: *Solid state physics literature guides*, 1999.

Glimm, J: *Quantification for Multiscale Simulations*, 2002.

Hakhoe, M: *Special issue on theoretical solid state physics*, 1995.

Harold, T: *Solid State Physics*, 1999.

Henricus, P: *Solid state physics: Magnetic properties of metals*, 1995.

Kachhava, C.M: *Solid State Physics*, 1998.

Kakurai, K: *Solid state physics: Magnetic properties of non-metallic inorganic*, 1999.

Klose, W: *Solid state physics: Superconductors, transition temperatures*, 1998.

Low, W: *Solid state physics: Advances in research and applications*, 2000.

Michiels, L: *Solid state physics in electronics and telecommunications*, 1995.

Paul, W: *Solid state physics research at Purdue*, 1999.

Schmid, D: *Solid-state physics*, 2001.

Thomas, P: *Solid state physics*, 1996.

Toroczkai, Z: *Continuous Extension of the Geometric Control Method*, 2001.

Treusch, J: *Advances in solid state physics*, 1994.

Vivian, A: *Solid state physics: part A: Preparation, structure, mechanical*, 1996.

Index